Darwin's Conservatives

Darwin's Conservatives

The Misguided Quest

John G. West

Description

Conservatives such as George Will, James Q. Wilson, and Larry Arnhart have mounted a vigorous defense of Darwinian biology, even urging conservatives to draw on Darwin's theory for support. In this small but incisive book, Dr. John West argues that the quest for "Darwinian conservatism" is misguided and fundamentally flawed. Contrary to claims by Darwin's conservatives, Darwinian evolution promotes relativism rather than traditional morality. It fosters utopianism rather than limited government. It is corrosive, rather than supportive, of free will and religious belief. Finally, and most importantly, Darwinian evolution is in tension with the scientific evidence.

Author Description

Dr. John G. West is a Senior Fellow of the Discovery Institute. Formerly the chair of the Department of Political Science and Geography at Seattle Pacific University, Dr. West has written, edited, or co-authored ten books, including *Traipsing into Evolution: Intelligent Design and the* Kitzmiller vs. Dover *Decision* and *The Politics of Revelation and Reason.*

Copyright Notice

Publisher's Note

This book is part of a series published by the Center for Science & Culture at Discovery Institute in Seattle. Previous books in that series include: *Traipsing into Evolution: Intelligent Design and the Kitzmiller v. Dover Decision, Are We Spiritual Machines?: Ray Kurzweil vs. The Critics of Strong A.I.* by Jay W. Richards et. al., *Getting the Facts Straight: A Viewer's Guide to PBS's Evolution* by the Discovery Institute, and *Why Is a Fly Not a Horse?* by Giuseppe Sermonti.

Library Cataloging Data

Darwin's Conservatives: The Misguided Quest

John G. West (1964–)

160 pages, 6 x 9 x 0.37 in. & 0.54 pound; 229 x 152 x 9.4 mm. & 245 grams.

1. Conservatism. 2. Social Darwinism--United States. 3. Ethics, Evolutionary.

4. Evolution–Religious aspects. 5. Darwin, Charles, 1809-1882–Influence

BISAC Subject Headings: POL042020—POLITICAL SCIENCE/Political Ideologies/ Conservatism & Liberalism. SCI075000—SCIENCE/Philosophy & Social Aspects. PHI005000—PHILOSOPHY/Ethics & Moral Philosophy. PHI019000—PHILOSOPHY/ Political. REL106000—RELIGION/Religion & Science.

ISBN-10: 0-9790141-0-7 ISBN-13: 978-0-9790141-0-9

Publisher Information

Discovery Institute Press, Discovery Institute. Internet: http://www. discovery.org/

Published in the United States of America on acid-free paper.

First Edition, First Printing. October 2006.

To Bruce Chapman, mentor and friend.

Acknowledgments

As this book attests, Larry Arnhart and I disagree on a number of things. Yet I would like to thank him for making this book possible. Prof. Arnhart is a defender of Darwin's theory who enjoys open and courteous debate, and I appreciate the opportunity to interact with him. When he published *Darwinian Conservatism* in 2005, he graciously sent me a copy and encouraged me to read it. Prof. Arnhart then invited me to participate in a panel exploring the themes of his book sponsored by the Claremont Institute at the American Political Science Assocation annual meeting in 2006. This small volume grew out of my presentation at that panel. It also draws on a much larger book project dealing with the impact of Darwinian reductionism and scientific materialism on American public policy during the last century, a project I hope will see the light of day sometime soon. In addition to Prof. Arnhart, I would like to thank Discovery Institute President Bruce Chapman, whose perceptive comments and suggestions improved several parts of the manuscript; graphic designer Brian Gage, who produced another one of his wonderful covers for the book; Mike Perry who typset and copy-edited the text and prepared the index; and Rob Crowther, who oversees the operation and marketing of Discovery Institute Press.

CONTENTS

INTRODUCTION

The debate over Darwinian evolution is usually framed by the news-media as a clash between "right" and "left." Conservatives are presumed to be critical of Darwin's theory, while liberals are presumed to be supportive of it.

As in most cases, reality is more complicated.

There always have been liberal critics of Darwin. In the early twentieth century, progressive reformer William Jennings Bryan fought for women's suffrage, world peace—and against Darwinism. More recently, left-wing novelist Kurt Vonnegut, a self-described "secular humanist," has called our human bodies "miracles of design" and faulted scientists for "pretending they have the answer as [to] how we got this way when natural selection couldn't possibly have produced such machines."[1]

Just as there have been critics of Darwin on the left, there continue to be champions of Darwinism on the right. In the last few years, pundits such as George Will, Charles Krauthammer, and John Derbyshire, along with social scientist James Q. Wilson and political theorist Larry Arnhart, have strongly defended Darwin's theory and denounced Darwin's critics.[2]

According to Will, "evolution" is a "fact," and anyone who does not recognize this elementary truth endangers the "conservative coalition."[3] After the Kansas State Board of Education called for students to hear the scientific evidence for and against Darwin's theory, Will castigated board members for being "the kind of conservatives who make conservatism repulsive to temperate people." Charles Krauthammer has likewise berated proponents of intelligent design for perpetuating scientific "fraud," and James Q. Wilson, writing for *The Wall Street Journal,* has insisted that "[t]he theory of evolution... is literally the only scientific defensible theory of the origin of species."[4]

Some of Darwin's conservatives even promote Darwinian biology as a way to defend conservatism. In his book *The Moral Sense,* James Q. Wilson draws on Darwinian biology to support traditional morality,[5] and writing in *National Review,* law professor John O. McGinnis has championed Darwinian sociobiology as a counter to left-wing utopianism. McGinnis opines that the future success of conservatism depends on evolutionary biology: "any political movement that hopes to be successful must come to terms with the second rise of Darwinism."[6]

No one has been more articulate in championing "Darwinian conservatism" than professor Larry Arnhart of Northern Illinois University, who argues that "[c]onservatives need Charles Darwin... because a Darwinian science of human nature supports conservatives in their realist view of human imperfectibility and their commitment to ordered liberty."[7] Like McGinnis, Arnhart suggests that conservatism may be doomed unless it embraces Darwinian biology. "The intellectual vitality of conservativism in the twenty-first century will depend on the success of conservatives in appealing to advances in the biology of human nature as confirming conservative thought."

In his recent book *Darwinian Conservatism,* Arnhart offers multiple reasons why he thinks Darwinism supports conservatism, as well as responding to various objections to Darwin's theory raised by some conservatives. As there is significant overlap between some of the reasons and objections discussed by Arnhart, I am going to group them into what I think are his seven main arguments: (1) Darwinism supports traditional morality; (2) Darwinism supports the traditional view of family life and sexuality; (3) Darwinism is compatible with free will and personal responsibility; (4) Darwinism supports economic liberty; (5) Darwinism supports "non-utopian limited government."; (6) Darwinism is compatible with religion; and (7) Darwinism has not been refuted by intelligent design.

Analyzing each of these arguments in turn, this book will argue that the quest to found conservatism on Darwinian biology is misguided and fundamentally flawed. Contrary to its conservative champions, Darwin's theory manifestly does not reinforce the teachings of conservatism. It promotes moral relativism rather than traditional morality. It fosters utopianism rather than limited government. It is corrosive, rather than supportive, of both free will and religious belief. Finally, and most importantly, Darwinian evolution is in tension with the scientific evidence, and conservatism cannot hope to strengthen itself by relying on Darwinism's increasingly shaky empirical foundations.

Defining "Darwinian Conservatism"

Before analyzing the case for Darwinian conservatism, we first need to define what we mean by both "conservatism" and "Darwinism."

According to Larry Arnhart, the core of conservatism is its "realist vision of human imperfectibility," which he sharply distinguishes from the left's "utopian vision of human perfectibility under the rule of abstract reason" as championed by the French Revolution.[8] Because of its realism, conservatism cherishes economic liberty, private property, limited government, and the restraints offered by tradition and religion. It also seeks to ground morality in the unchanging truths of human nature. Arnhart's description of conservatism may not be exhaustive, but it provides a clear working definition.

The same cannot be said for Arnhart's use of the term "Darwinian" and "Darwinism." While Arnhart painstakingly discusses what he means by conservatism, he provides surprisingly little clarity about the limits of his definition of "Darwinism." Sometimes Arnhart uses the term to refer to natural selection and random variation in nature.[9] Other times he appears to apply it to any and every biological insight into human nature.[10] Still other times he applies the term outside of biology to cultural evolution.[11] Underlying the ambiguous usage is the slippery way the term "evolution" itself is often employed in contemporary discourse. As has been noted by others, the term "evolution" has multiple meanings, including change over time, descent with modification, incremental changes within species or gene pools ("microevolution"), large-scale changes and the development of fundamentally new biological organisms or features ("macroevolution"), and Darwin's undirected mechanism of natural selection and random variation.[12] Given the ambiguity of terms in this field, it is critical that we be as clear as possible about what we are talking about when we invoke the phrase "*Darwinian* conservatism."

Unfortunately, Arnhart and other "Darwinian conservatives" often seem to conflate any attempt to understand human beings biologically with "Darwinism." Thus, Arnhart repeatedly cites research showing that various human behaviors or traits have a biological basis. But how are these research findings unique to "Darwinism"? After all, one can believe that there is a biological basis for, say, gender differences, without believing in the Darwinian story about how those gender differences first arose. Darwinism purports to explain the historical process by which a biological trait came to be. But the empirical observation that a biological trait currently exists owes little or nothing to Darwinian theory. We can learn that a certain gene is connected to a certain biological trait without knowing anything about how that gene first developed.

In fact, Darwinian theory is surprisingly irrelevant to much of modern biological research. Despite claims by some Darwinists that "nothing in biology makes sense except in the light of evolution,"[13] little of our understanding about how organisms currently operate is truly dependant on Darwinian theory. According to National Academy of Sciences member Philip Skell, Darwinian theory is largely "superfluous" in biological explanations of how things work.[14] Skell, the Emeritus Evan Pugh Professor at Pennsylvania State University, argues that "Darwinian evolution... does not provide a fruitful heuristic in experimental biology" and "the claim that it is the cornerstone of modern experimental biology will be met with quiet skepticism from a growing number of scientists in fields where theories actually do serve as cornerstones for tangible breakthroughs." Skell reports that he "recently asked more than 70 eminent researchers if they would have done their work differently if they had thought Darwin's theory was wrong. The responses were all the same: No." Moreover, when Skell examined "the outstanding biodiscoveries of the past century" such as "the discovery of the double helix" and the "mapping of the genomes," he "found that Darwin's theory had provided no discernible guidance, but was brought in, after the breakthroughs, as an interesting narrative gloss."

All of this is to say that the debate over Darwinism is *not* primarily a debate between those who believe there are biological bases of human behavior and those who do not. One can accept the idea that certain human traits have a biological basis without ever embracing the speculative causal history offered by Darwinists to explain the origins of the trait.

So when I use the term "Darwinian evolution" or "neo-Darwinism" or just simply "Darwinism," I am *not* merely referring to a biological understanding of nature, or even to an evolutionary understanding of nature. Instead, I am referring to the two major claims made by the modern theory of evolution known as "neo-Darwinism." First, all living things developed through an evolutionary process of descent with modification from a universal common ancestor. Second, the primary mechanism of evolution is an unguided material process of natural selection acting on random variations—today understood as random mutations. The second claim is by far the most crucial: It is what makes the modern theory of evolution *Darwinian*. Plenty of people believed in an evolutionary history of life prior to Darwin. Darwin's key contribution was offering a seemingly convincing account of the mechanism of evolution.

It is important to emphasize that the Darwinian mechanism is non-teleological. Darwin thought he had explained how we can get the appearance of design throughout nature through a process that did not have the design of particular organisms or biological structures in mind.

The only "purpose" of natural selection, if we can call it that, is immediate survival. Organisms best adapted to their current environment are more likely to survive and reproduce than organisms less adapted to their current environment. Natural selection is blind to the future, and thus in no sense are particular organisms or biological structures to be considered the purposeful result of evolution. In the Darwinian view, biological structures such as the vertebrate eye, or the wings of butterflies, or the bacterial flagellum, developed through the interplay of chance (random mutations) and necessity (natural selection or "survival of the fittest"). The same holds true for the higher animals such as human beings. In the words of Harvard paleontologist George Gaylord Simpson, "[m]an is the result of a purposeless and natural process that did not have him in mind."[15]

It must be emphasized further that this rejection of teleology is fundamental rather than incidental to Darwin's theory. Of course, evolution *can* be conceived of as a purposeful process, but then it would no longer be *Darwinian* evolution. A belief in the nonteleological character of evolution is not just embraced by a few militants, such as the much derided whipping boy for atheistic evolution, Richard Dawkins. It has been the dominant view of modern evolutionary biologists, and it was Darwin's own view. Criticizing those who believed that evolution was somehow guided, Darwin wrote:

> no shadow of reason can be assigned for the belief that variations... which have been the groundwork through natural selection of the formation of the most perfectly adapted animals in the world, man included, were intentionally and specially guided. However much we may wish it, we can hardly follow Professor Asa Gray in his belief "that variation has been led along certain beneficial lines," like a stream "along definite and useful lines of irrigation."[16]

The insistence that evolution acts without plan or purpose has been a standard refrain in science textbooks. According to the college biology text *A View of Life* (1981), Darwinian evolution is "a natural process without purpose or inherent direction."[17] According to *Biology* (1998), "[e]volution is random and undirected" and "works without either plan or purpose."[18] According to *Evolutionary Biology* (1998), "[b]y coupling undirected, purposeless variation to the blind, uncaring process of natural selection, Darwin made theological or spiritual explanations of the life processes superfluous."[19] The "profound, and deeply unsettling, implication" of this "purely mechanical, material explanation for the existence and characteristics of diverse organisms is that *we need not invoke, nor can we find any evidence for, any design, goal, or purpose anywhere in the natural world,* except in human behavior."[20] According to *Life: The Science of Biology* (2001), accepting "the Darwinian view.... means accepting not only the processes of evolution, but also the view that... evolutionary change occurs without any 'goals.' The idea that evolutionary change is not directed toward a final goal or state has been more difficult for many people to accept than the process of evolution itself."[21]

Despite the plenitude of such explicit statements, the purposeless character of Darwinian evolution is often obscured in current debates. The reason for this is preeminently public relations rather than science. Astute defenders of Darwin's theory understand that publicly insisting on its blind and unguided character will sound anti-religious and even incredible to most people. So they resort to self-censorship and artful rewording in order to neutralize public opinion.

Consider the controversy that ensued after the National Association of Biology Teachers (NABT) adopted a statement defining evolution as "an unsupervised, impersonal, unpredictable and natural process," and stressing that natural selection "has no specific direction or goal, including survival of a species."[22] The admirable clarity of the NABT definition quickly made it a lightning rod for criticisms by Darwin's critics. Seeking to defuse what she called "an unanticipated public relations problem,"[23] Eugenie Scott of the National Center for Science Education (NCSE) successfully lobbied the NABT board to excise the words "unsupervised" and "impersonal" from the definition.[24] Revealingly, Scott insisted that the change dropped "two nonessential words" and "would not affect the statement's accurate characterization of evolution."[25] She emphasized in particular that evolution was still described as a "natural process" and natural selection was still defined as having "no specific direction or goal."[26] In other words, the rewording adopted by the NABT solved the "public relations problem" without changing the meaning of the NABT statement.

Scott, a self-described "evolution evangelist,"[27] understands the need to downplay Darwinism's denial of purpose in public, and she does her best to convince her fellow evolutionists to act accordingly. It is a challenging task. It can be difficult for scientists to talk about the purposeless nature of evolution when interacting with their colleagues, and then suddenly to switch gears and suppress such comments when talking to the public. After the NABT revised its definition of evolution, more than 100 evolutionists signed an open letter decrying the change as "the first wedge of a movement intended to surreptitiously introduce religious teachings into our public schools."[28] According to the signers of the open letter, there was nothing wrong with the words "unsupervised" and "impersonal" because "evolution indeed is, to the best of our knowledge, an impersonal and unsupervised process." According to the signers, the NABT's artful change was disingenuous because it "leaves open the possibility that evolution is in fact supervised in a personal manner. This is a prospect that every evolutionary biologist should vigorously and positively deny."

The difficulty of getting evolutionists to read from the same public relations script can sometimes lead to amusingly contradictory messages for public consumption, as it did last year in the state of Kansas. In 2005 Kansas adopted new science standards that defined "biological evolution" as "descent with modification," but added that evolution "postulates an unguided natural process that has no discernable direction or goal."[29] This wording was clearly inspired by the standard NABT definition of evolution, but Kansas evolutionists attacked it, insisting that the language wrongly implied that modern evolutionary theory is anti-religious. Lo and behold, a group of 38 Nobel laureates then sent an open letter to the Kansas Board of Education denouncing its new standards on evolution.[30] But someone obviously had failed to send them the right talking points. In their letter, the Nobel laureates defined evolution as "the result of an *unguided, unplanned* process of random variation and natural selection."[31] In other words, they essentially embraced the very definition that local evolutionists had denounced the Board of Education for adopting!

Having clarified the meaning of both "conservatism" and "Darwinism," we are now ready to scrutinize the case for "Darwinian conservatism."

CHAPTER I

DOES DARWINISM SUPPORT TRADITIONAL MORALITY?

"Morality... is merely an adaptation put in place to further our reproductive ends.... In an important sense, ethics as we understand it is an illusion fobbed off on us by our genes to get us to cooperate."
MICHAEL RUSE AND E. O. WILSON, "THE EVOLUTION OF ETHICS."[32]

The question of whether Darwinian evolution supports traditional morality is an old one. In a famous essay on "Evolution and Ethics," Darwin's "bulldog" Thomas Huxley vigorously argued in the negative: "the practice of that which is ethically best—what we call goodness or virtue—involves a course of conduct which, in all respects, is opposed to that which leads to success in the cosmic struggle for existence." [33] In Huxley's view, "the ethical progress of society depends, not on imitating the cosmic process [of evolution]... but in combatting it."[34]

Darwinian conservatives such as James Q. Wilson and Larry Arnhart beg to differ. In his book *The Moral Sense*, Wilson draws on Darwin's evolutionary account of morality in *The Descent of Man* to reinforce the conclusions of traditional morality.[35] In his book *Darwinian Conservatism*, Arnhart likewise insists that "rather than assuming that Darwinism subverts morality, conservatives should recognize that a Darwinian view of human nature reinforces the conservative concern for cultivating moral character."[36] Indeed, Arnhart seeks to persuade conservatives that Darwinism can be enlisted to refute "cultural relativism." "As far as I can see," he writes, "the only escape from such cultural relativism is to argue that there is a universal human nature of natural instincts and desires," and Darwinism supplies the scientific proof of that universal nature, enabling us to "judge some societies... as satisfying those natural desires more fully than other societies."[37]

To justify these claims, Arnhart—like Wilson—highlights Darwin's belief that human beings have a natural "moral sense" that instructs them about right and wrong. He argues that Darwin's "moral sense" provides a biological grounding for the teachings of traditional morality. There is something to be said for this argument. Darwin claimed that man's social instincts arise out of his biology, and that these social instincts "naturally lead to the golden rule, 'As ye would that men should do to you, do ye to them likewise.'" According to Darwin, this maxim "lies at the foundation of morality."[38] According to Darwin, then, it would seem that evolution ultimately promotes the morality of Jesus rather than the law of the jungle.

However, before one rushes to install Darwin in the pantheon of defenders of a natural moral law, one needs to scrutinize the account of morality Darwin gave prior to his conclusion supporting conventional morality. While according to Darwin nature has led to the golden rule, it did not do so because the golden rule is somehow intrinsically right. It did so because the golden rule ultimately is connected to self-preservation. At the base of the golden rule in Darwin's view are the social instincts, and these developed primarily because they promote survival: "Those communities, which included the greatest number of the most sympathetic members, would flourish best and rear the greatest number of offspring."[39] In the conclusion to *The Descent of Man*, Darwin made this point even more clearly, stating that "the... origin of the moral sense lies in the social instincts, including sympathy; and these instincts no doubt were primarily gained, as in the case of the lower animals, through natural selection."[40]

But what happens in cases where traditional morality does *not* happen to promote survival? If human morality is ultimately grounded in the struggle to survive, it seems optimistic in the extreme to think that the by-product will always be something akin to traditional Judeo-Christian morality. Darwin himself provided exhaustive evidence on this point. While he tried to show that traditional virtues such as courage and love were products of nature, he also demonstrated that a great many vices are no less a product of the struggle for existence. Maternal instinct is natural, but so is infanticide.[41] Care toward family members is natural, but so is euthanasia of the feeble, even if they happen to be one's parents:

> That animals sometimes are far from feeling any sympathy is too certain; for they will expel a wounded animal from the herd, or gore or worry it to death. This is almost the blackest fact in natural history, unless indeed the explanation which has been suggested is true, that their instinct or reason leads them to expel an injured companion, lest beasts of prey, including man, should be tempted to follow the troop. In this case their conduct is not much worse than that of the North American Indians who leave their feeble comrades to perish on the plains, or the Feegeans, who, when their parents get old or fall ill, bury them alive.[42]

Throughout his discussion of morality, Darwin repeatedly referred to "higher" and "lower" moral impulses as if there were some transcendent standard of morality to which he compared human and animal behavior. Darwin wrote as if conventional virtues such as kindness and courage were objectively preferable to conventional vices such as cruelty and lust. But it is difficult to make sense of such comments in terms of Darwin's own system, which clearly portrayed morality as ultimately reducible to that which promotes biological survival.

In the current set of circumstances Darwin could believe that his view meant the extension of benevolence "to the men of all races, to the imbecile, the maimed, and other useless members of society, and finally to the lower animals."[43] He could even hope that "looking to future generations, there is no cause to fear that the social instincts will grow weaker, and we may expect that virtuous habits will grow stronger, becoming perhaps fixed by inheritance... and virtue will be triumphant."[44]

But even Darwin would have to acknowledge, if pressed, that given a different set of circumstances, a radically different conception of morality might be dictated. At one point, he said as much: "If, for instance... men were reared under precisely the same conditions as hive-bees, there can hardly be a doubt that our unmarried females would, like the worker-bees, think it a sacred duty to kill their brothers, and mothers would strive to kill their fertile daughters; and no one would think of interfering."[45] Although this passage references the behavior of hive bees, it is making a point about *human* morality and how it is ultimately a function of the conditions of survival. Whenever those conditions change, Darwin seems to say, so too will the maxims of human morality.

Temperate and considerate in his own life, Darwin may have sincerely believed that his biology supported traditional morality.[46] Nevertheless, the internal logic of his theory did not allow any permanent foundation for ethics other than the struggle to survive, and for that reason his attempt to square a biological understanding of ethics with traditional morality is ultimately unpersuasive. Far from defending a traditional conception of morality, Darwin's account of the moral sense supplies a basis for the very relativism that Arnhart fears.

I would like to add here that I am not quarreling with Arnhart's attempt to enlist biology to support traditional morality. I actually agree with him that showing a biological basis for certain moral desires could conceivably reinforce traditional morality—*but only if we have reason to assume that those biological desires are somehow normative*. What Darwinian conservatives fail to appreciate is how the Darwinian account of the origins of biological traits fundamentally undercuts their effort to treat our biological desires as normative. If one believes that natural desires have been implanted in human beings by intelligent design, or even that they represent irreducible and unchanging truths inherent in the universe, it would be rational to accept those desires as a grounding for a universal code of morality. But Darwinism explicitly denies that natural desires are either the result of intelligent design or an unchanging nature.

Arnhart acknowledges that human morality needs to be grounded in a human nature that he variously describes as "universal," "unchanging," or at least "enduring," in order to avoid the twin challenges of relativism and utopianism.[47] The problem is that Darwinian theory does not allow for such a human nature.[48] According to Darwinism, human nature forever evolves according to the immediate dictates of the environment. Whenever the environment changes, human nature has the potential for changing too. As previously discussed, the ultimate basis for these changes to human nature is self-preservation or physical survival. According to Darwin, specific moral precepts develop merely because under certain environmental conditions they promote survival. But once the conditions for survival change, so too do the dictates of morality. That is why we find in nature both the maternal instinct and infanticide, both courage and cowardice, both honor and treachery. In short, natural selection "chooses" whatever traits best promote survival under the existing circumstances. Sometimes that may include traits we consider "moral," but other times it will include shocking immoralities. This is the natural morality of Hobbes and Machiavelli, not Aristotle and Aquinas.

It should be pointed out that the Darwinian view makes it very difficult to condemn as evil any human behavior that has persisted among human beings, because every trait that continues to exist even among a subpopulation has an equal right to claim nature's sanction. Presumably even anti-social behaviors such as fraud and pedophilia must continue to exist among human beings because they were favored at some point by natural selection and therefore have some sort of biological basis. While we can say that desires to perpetrate fraud or engage in pedophilia are not right for those of us who do not have those desires, on what basis do we morally condemn those who *do* hold such desires? Their desires were implanted in them by natural selection in the same way that our desires were implanted in us. And who are we to condemn what nature has sanctioned? As a matter of self-preservation, we certainly can try to stop such behaviors by force if we have sufficient power to do so, but that is not the same thing as being able to condemn pedophiles or tax cheats as morally blameworthy.

According to the Darwinian conception of morality, the rules of conventional morality can only be said to be obligatory on those whom natural selection has granted a conventional moral sense. This becomes clear in Arnhart's extended discussion of psychopaths in his book *Darwinian Natural Right,* where Arnhart says "[w]e cannot properly blame psychopaths for lacking the moral sentiments natural to us" and concedes that "psychopaths are under no moral obligation to conform to the moral sense, because they lack the moral emotions that provide the only basis for moral obligation...."[49] Thus, according to Arnhart, a particular moral obligation exists only insofar as natural selection has implanted that obligation in a specific person.[50]

But this means that anyone programmed by natural selection with desires contrary to traditional morality cannot be judged by traditional morality. "By the logic of the case," writes J. Budziszewski, "*everyone* whose desires are significantly different than the rest of us gets his own morality."[51] Does this result not encourage the very relativism that Arnhart seeks to oppose? Of course, Arnhart says with regard to psychopaths that we can still restrain them by force as a matter of our self-preservation.[52] True, but that is a far cry from Arnhart's original proposal of using biology to combat relativism. It is also a far cry from traditional natural law theory, which provides a basis for condemning as objectively wrong any violation of the moral law.

CHAPTER II

Does Darwinism Support the Traditional Family?

"We cannot escape our animal origins."
Malcolm Potts and Roger Short, *Ever Since Adam and Eve: The Evolution of Human Sexuality*[53]

Darwinian conservatives extend their biological defense of morality into one of the most difficult and divisive cultural battlegrounds for conservatives—family life and sexuality. Larry Arnhart, for one, is convinced that Darwinism can be used to defend the traditional family:

> a Darwinian account of the natural desires for sexual identity, sexual mating, and parental care confirms the conservative commitment to the traditional social order of sex, marriage, and the family. While those on the left tend to see sexual differences, family life, and parental care as social constructions that can be changed—and perhaps even abolished—by social engineering, Darwinian biology sustains the conservative understanding of sexual conduct and familial bonding as innate propensities of human nature.[54]

But the problem of Darwinian relativism discussed earlier applies here as well. Monogamy is natural according to Darwinism, but so is adultery. Marital fidelity is natural, but so is promiscuity. Parental love is natural, but so is infanticide. Since Darwinism provides no basis for preferring one natural trait over another, we are left with a biological justification for sexual relativism rather than the traditional family.

Darwinism's inherent relativism is apparent in much of the writing by Darwinists on sex and family life over the past century, although one would never ascertain this fact from reading Arnhart. Consider his discussion of Darwin's account of marriage in *The Descent of Man*. Arnhart depicts Darwin as critical of anthropologists who "believed that marriage and family life were not natural because originally primitive human beings were completely promiscuous in their sexual intercourse and thus there were no enduring marital or familial ties."[55] Arnhart quotes Darwin declaring that he "cannot believe that absolutely promiscuous intercourse prevailed in times past" because, in Arnhart's words, "the sexual jealousy of males and the instinctive tie between mother and child would naturally favor some kind of sexual pair-bonding and parent-child bond."[56]

But Arnhart presents a remarkably sanitized version of Darwin's evolutionary understanding of marriage. Significantly, he fails to quote the sentence in *The Descent of Man* immediately preceding Darwin's expression of doubt that the earliest humans were "absolutely promiscuous." In that earlier sentence, Darwin declared: "it seems probable that the habit of marriage, in any strict sense of the word, has been gradually developed; and that *almost promiscuous* or *very loose* intercourse was once extremely common throughout the world."[57] So while Darwin may have doubted that early humans were "absolutely promiscuous," he agreed that they practiced "*almost promiscuous* or *very loose* intercourse." At the same time, Darwin made clear that what we call marriage ("in the strict sense of the word") was *not* the original form of human mating practice, but "has been gradually developed" through the process of natural selection. Rejecting the Judeo-Christian conception of marriage as an institution ordained from the inception of humanity, Darwin depicted marriage as yet another evolving product of the struggle to survive.

What did Darwin mean by maintaining that early humans were "almost promiscuous" rather than "absolutely promiscuous"? As Arnhart correctly explains, Darwin believed that our instincts "naturally favor some kind of sexual pair-bonding and parent-child bond."[58] The key phrase here, which Arnhart does not elucidate, is "some kind." The type of bonding Darwin believed was favored by nature is rather different from what most conservatives would understand as traditional marriage. In the first edition of *The Descent of Man* Darwin appeared to conclude that the original pattern of human sexual bonding was some form of polygamy: "the most probable view is that primeval man aboriginally lived in small communities, each with as many wives as he could support and obtain."[59] While establishing a biological basis for polygamy as the original form of human marriage may give comfort to modern polygamists, I sincerely doubt many conservatives would find it useful to their efforts to defend marriage. Polygamy, after all, is not what most conservatives mean by "traditional marriage."

It is true that Darwin reworded this statement in a later edition to make it more open to monogamy, emphasizing that primeval men who lacked power could have been limited to "a single wife."[60] Yet before one cites Darwin's revised statement as a ringing endorsement of monogamy, it is important to understand how flexibly Darwin defined "monogamy" in *The Descent of Man*. In his view, monogamy did not necessarily mean lifelong pairing. "The pairing may not last for life," he wrote, " but only for each birth."[61] According to this understanding, the welfare mother who has five children fathered by three different men would still be practicing monogamy. Last time I checked, this is not the kind of monogamy most conservatives are seeking to defend.

An even more frank expression of Darwinian relativism toward marriage and sexuality can be found in the writings of Finnish sociologist Edward Westermarck (1862-1939), whose *History of Human Marriage* tried to provide a comprehensive account of the origin and development of human mating practices.[62] Incredibly, Arnhart cites Westermarck as a defender of the universality of marriage without discussing Westermarck's explicit embrace of moral relativism.

Inspired by Darwin's *Descent of Man*, Westermark believed that marriage, as well as other human sexual behaviors, had developed through natural selection.[63] Arnhart is right that insofar as marriage was an institution grounded in biology, Westermarck was critical of efforts by radical sex reformers to completely abolish it.[64] He viewed it for the foreseeable future as a permanent feature of human society. At the same time, however, Westermarck embraced moral relativism, arguing that "the emotional origin of moral judgments" refuted "the objective validity ascribed to them both by common sense and by normative theories of ethics."[65]

Applying his relativized understanding of morality to sexuality, Westermarck cast doubt on the validity of Judeo-Christian sex taboos, including homosexuality and even bestiality.[66] In *The Future of Marriage*, for example, Westermarck contended that "the moral condemnation of bestiality" was irrational because it was based on "an entirely emotional foundation.... It is therefore an opinion which is nowadays gaining ground that it should not be punished at all." To prove his point, Westermarck related the views of "a group of mountaineers" in Morocco who had told him that "a person who has intercourse with another man's animal has to buy for it new shoes, a new pack-saddle" and various other things. But when "asked what would happen if the animal was his own, the answer was, amidst much laughter, 'Why should not a man be allowed to do with his animal whatever he likes?'"[67]

Westermarck did not see biology as preserving traditional Judeo-Christian sexual morality. Indeed, he predicted "that in questions of sex people [in the future] will be less tied by conventional rules and... will recognise greater freedom for men and women to mould their own amatory life."[68]

Darwin and Westermarck are models of restraint compared to the Darwinist who did the most to shape our contemporary attitudes toward sex. Curiously, this Darwinist does not even merit a mention in either of Arnhart's two books on Darwinism. Yet his name has become a household word among educated Americans: Alfred Kinsey. Kinsey was an evolutionary biologist who started out studying gall wasps.[69] He then applied the reductionist framework he learned from evolutionary biology to the study of human sexuality in the 1940s and 50s. Kinsey is a sobering example of a thinker who pressed evolutionary reductionism to its logical conclusion.

Kinsey treated the "human animal" as merely another type of mammal whose mating behavior could be fully explicated in terms of biology and conditioning. His ultimate goal was to refashion human sexual morality according to the standard of normal mammalian behavior. Time and again, he argued that sex behaviors traditionally proscribed by society were widespread in the animal kingdom as well as parts of human society, and therefore efforts to punish or repress those activities were unnatural as well as futile.

The taboo against bestiality, for example, contradicted an "increasing number" of reports of "higher mammals... mating, or trying to mate, with individuals of totally distinct and sometimes quite remote species," as well as research findings that a substantial proportion of "rural boys" had "sexual contacts with animals to the point of orgasm."[70] Accordingly, Kinsey suggested that opposition to bestiality originated in "superstition"[71] and encouraged doctors and psychologists to assure young men who engaged in the practice of the normality of their behavior.[72] Sex behavior among children was likewise natural according to Kinsey,[73] as was male adultery and promiscuity: "There seems to be no question but that the human male would be promiscuous in his choice of sexual partner throughout the whole of his life if there were no social restrictions. This is the history of his anthropoid ancestors, and this is the history of unrestrained human males everywhere."[74]

Perhaps Arnhart would claim that Kinsey was not a genuine Darwinist because he seemed to place greater emphasis on environmental conditioning than on heredity in determining human sexual practices. In truth, Darwinism relies on both environment and heredity, and Darwin himself emphasized the importance of the environment in the development of human culture. Genetic reductionism is one form of Darwinism, but environmental reductionism is surely its handmaiden. The fact that Kinsey ascribed more of our sexual practices to environmental conditioning than some other evolutionary biologists does not disqualify him as a Darwinist. More importantly, Kinsey's recognition that any widespread behavior among animals and humans must be just as "natural" as any other widespread behavior represents a logical culmination of the Darwinian approach.

Post-Kinsey, the tendency toward Darwinian sexual relativism has continued even as hereditarian accounts of human behavior have returned to favor through "sociobiology" and "evolutionary psychology," which purport to offer evolutionary explanations for the full range of human sexual practices. According to Randy Thornhill and Craig Palmer, for example, "[t]he ultimate causes of human rape are clearly to be found in the distinctive evolution of male and female sexuality."[75] The same is true for casual sex and extramarital affairs. According to evolutionary psychology proponent Robert Wright, Darwin's theory explains why husbands are much more likely to desert their wives than vice versa: "The husband can, in principle, find an eighteen-year-old woman with twenty-five years of reproduction ahead. The wife... cannot possibly find a mate who will give her twenty-five years worth of reproductive potential."[76] An evolutionary psychology textbook, meanwhile, claims that casual sex is an evolutionary adaptation based not only on "obvious reproductive advantages... to men" but also "tremendous benefits to women."[77]

Proponents of evolutionary psychology typically offer caveats that just because natural selection programmed a certain behavior does not make it morally right. Yet these caveats cut against the logic of their own position. If fidelity and adultery both exist simply because they furthered the survival of the fittest genes, what objective basis do we have for preferring one trait over the other?

The impact of evolutionary psychology's approach to sexuality on popular culture could be seen during the scandal over President Bill Clinton's extramarital relationship with intern Monica Lewinsky. One newspaper article published during the controversy bore the headline:

"Not Meant for Monogamy? Blame the genes: Evolutionary psychologists and biologists suggest that humans are naturally polygynous, with perpetuating the species the goal—which means that sticking with one woman is not an efficient reproductive strategy."[78]

One reason Arnhart's account of Darwinian sexuality is less-than-persuasive is that he never grapples with this long history of Darwinian sexual relativism and reductionism. Indeed, reading Arnhart one might conclude that Darwinists have been the guardians of traditional morality in the bedroom over the past century, when in fact they have been on the vanguard of subverting traditional sexual ethics. Of course, it is logically possible that the Darwinian relativists have been wrong and that Arnhart—and James Q. Wilson, and a handful of other conservatives—have suddenly discovered the true meaning of a Darwinian approach to sexuality. But it does seem that an explanation is required as to why the leading Darwinian thinkers of the past century have thought otherwise. Were they all wrong in Arnhart's view? If so, why? Arnhart does not answer these questions, leaving his readers to wonder whether his account of Darwinian sexuality is merely idiosyncratic.

Be that as it may, it must be emphasized again that the entire effort to justify traditional morality using Darwinian biology is ill-conceived. Even if conservatives convinced Darwinian biologists and ethicists to abandon their relativism, there is no getting around the fact that in the Darwinian system there can be no one "naturally right" pattern of sexual relations between human beings. According to Darwinism, there is nothing sacred or permanent about any of the forms of family life found in nature. They are all adaptations to the particular environment humans faced, and presumably when the environment changes, so too will the adaptations.

This does not mean that we can learn nothing from modern biology about sex, family life, and gender. Research confirming innate differences between men and women, or neuro-chemical factors supporting parental attachment to their children, can corroborate ethical standards we already hold on other grounds. But the reductionist Darwinian account of how these traits first arose in nature adds nothing to the conservative case for the traditional family. After all, one can identify biological supports for human behavior without even broaching the question of how they were originally generated. At best, a Darwinian gloss on these biological traits is superfluous. At worst, it undermines the moral legitimacy of the very traits conservatives seek to defend by ascribing them to a purposeless material process whose only real goal is physical survival.

CHAPTER III

Is Darwinism Compatible with Free Will and Personal Responsibility?

"Naturalistic evolution has clear consequences that Charles Darwin understood perfectly.... [including the idea that] human free will is nonexistent.... Free will is a disastrous and mean social myth."
William Provine, Professor of History of Biology, Cornell University.[79]

In *The Descent of Man,* Darwin explained human behavior largely as the function of pre-determined—and often anti-social—instincts. For all of Darwin's praise of man's sociability, he wrote that "it cannot be maintained that the social instincts are ordinarily stronger in man, or have become stronger through long-continued habit, than the instincts... of self-preservation, hunger, lust, vengeance, &c."[80] What did this mean in practice? "At the moment of action," wrote Darwin, "man will no doubt be apt to follow the stronger impulse; and though this may occasionally prompt him to the noblest deeds, it will far more commonly lead him to gratify his own desires at the expense of other men."[81]

Darwin tried to soften the implications of his view by going on to claim that men will learn to regret their impulsive actions and eventually this regret will create in them a conscience. However, Darwin did not convincingly explain why the conscience would trump instincts he earlier depicted as so overwhelming. Even if conscience is able to counteract the anti-social instincts in some men, presumably those who act anti-socially are only following their own strongest instincts. If this be the case, how responsible are those who act against society?

Darwin in *The Descent of Man* doesn't directly address the consequences of his account for free will and personal responsibility. He was more open in his in unpublished notebooks. There he wrote that "the general delusion about free will [is] obvious," and that one ought to punish criminals "solely to *deter* others"—not because they did something blameworthy.[82] "This view should teach one profound humility," wrote Darwin, "one deserves no credit for anything... nor ought one to blame others." Darwin denied that such a fatalistic view would harm society because he thought that ordinary people would never be "*fully* convinced of its truth," and the enlightened few who did embrace it could be trusted.

There is no question that materialists have found inspiration in Darwin's view that man's mental faculties arose through a purely purposeless material process of chance and necessity. In the words of nineteenth-century German physiologist Emil Du Bois-Reymond, "the evolution theory in connection with the doctrine of natural selection forces upon [man]... the idea that the soul has arisen as the gradual result of certain material combinations."[83] Noted evolutionist Stephen Jay Gould expressed the same view, arguing that according to Darwin's theory "matter is the ground of all existence: mind, spirit, and God as well, are just words that express the wondrous results of neuronal complexity."[84]

It should be no surprise, then, that attacks on free will and personal responsibility have featured prominently in Darwinian accounts of human behavior during the past century-and-a-half. For example, Darwinism played a key role in the development of the "new school of criminology" by Cesare Lombroso and others in the late nineteenth century. These criminologists tried to find Darwinian explanations for why people engaged in crime, even labeling some persons "born criminals" because they were supposed to be throwbacks to an earlier stage in evolutionary history. Lombroso and his followers repudiated the traditional idea that "crime involved... moral guilt." Italian Jurist Enrico Ferri, one of Lombroso's most celebrated disciples, argued that it was no longer reasonable to believe that human beings could make choices outside the normal chain of material cause and effect given the advent of modern science, particularly the work of Charles Darwin. Ferri looked forward to the day when punishment and vengeance would be abandoned and crime would be treated as a "disease."[85]

The diminishment of free will is likewise rampant among today's purveyors of sociobiology and evolutionary psychology. MIT psychologist Steven Pinker, who Arnhart cites with approbation, has argued publicly for more lenient treatment of mothers who commit infanticide. Why? According to Pinker, natural selection made them do it! "[T]he emotional circuitry of mothers has evolved to cope with th[e] uncertain process [of raising children], so the baby killers turn out to be not moral monsters but nice, normal (and sometimes religious) young women."[86]

In his bestselling book *The Moral Animal*, evolutionary psychology booster Robert Wright goes even further, declaring "free will is an illusion, brought to us by evolution"[87] and "[u]nderstanding the often unconscious nature of genetic control is the first step toward understanding that —in many realms, not just sex—we're all puppets."[88] Wright does add that "our best hope for even partial liberation is to try to decipher the logic of the puppeteer."[89] But if "free will is an illusion," precisely how can we liberate ourselves from "the puppeteer"? And if human beings truly are "puppets" to their genes, puppets whose "emotions are just evolution's executioners"[90] (again quoting Wright), in what sense can people be blamed if they simply act according to their deepest impulses?

It is true that a number of Darwinists are likely repelled by the implications of their own theory when it comes to free will. Thus, while evolutionist William Provine at Cornell openly proclaims the denial of free will as a corollary of Darwinism, he concedes that "[e]ven evolutionists have trouble swallowing that implication."[91] The real question is not whether some evolutionists are squeamish about denying free will, but whether their scientific outlook allows them any rational basis to affirm it. Sociobiologist David Barash is more honest than many in admitting the tension between his own subjective experience of free will in daily life and his belief that "there can be no such thing as free will for the committed scientist."[92] Barash is willing to live with what he calls the "unspoken hypocrisy" of preaching materialistic determinism in public even while acting as if he has free will in private. At least he is willing to admit his hypocrisy. The point here is that Darwinists who try to cling to free will do so in spite of their theoretical commitment to materialism, not because of it.

Darwinian conservatives obviously want to do better than Barash and find a way to make Darwinism actually consistent with free will. Recognizing the debilitating impact of what he calls "strong reductionism," Arnhart does his best to disentangle Darwinism from it, insisting that "[i]n contrast to the reductionism often associated with modern science, Darwinian conservatism affirms the idea of emergence."[93] By "emergence," Arnhart means there are "special capacities of the human soul... manifesting the emergent complexity of life, in which higher levels of organization produce mental abilities that cannot be found at lower levels."[94] Whether "emergence" truly helps make Darwinism safe for free will, however, is doubtful.

To make his case, Arnhart draws on the work of Dr. Jeffrey Schwartz of UCLA, whose fascinating research seeks to demonstrate that our mental thoughts can produce physical changes in the brain. For Arnhart, the clear lesson of Schwartz's research is that "the mind that emerges from the human brain can change the brain itself. This emergent power of the brain for mental attention is the natural ground for human freedom."[95] Yet it is not clear what the word "emergent" adds to Arnhart's description. Schwartz's research does try to show the power of the human mind to act on the physical brain. But in and of itself it does not establish *how the power of the mind first developed*—whether it emerged from a purely purposeless material process, as Arnhart contends, or through a purposeful process directed by a preexisting intelligence, as has been more traditionally believed. Nor does Schwartz's research demonstrate whether the human mind is purely material (but "emergent") or the fusion of matter with a nonmaterial entity. Again, the focus of Schwartz's research is to show that the mind is real by demonstrating its effects on the brain, not to decide the debate over emergence.

Additionally, it is ironic that Arnhart would rely on the work of Schwartz at all, because Schwartz's research did not spring from Darwinian biology. In fact, Schwartz is openly critical of neo-Darwinism and supportive of intelligent design, and he is affiliated with a pro-intelligent design professional society established by mathematician and philosopher William Dembski, one of intelligent design's most prominent proponents.[96] If Darwinism is so compatible with emergence, why couldn't Arnhart cite research done by a committed Darwinist to establish his idea of emergence? Why is the most convincing research he could find being conducted by a critic of Darwinism?

Arnhart's championing of "emergence" notwithstanding, the history of Darwinian explanations of human behavior during the past century has been overwhelmingly a history of reductionism. And although Arnhart claims that "conservatives have no reason to fear a Darwinian science of human life as promoting a reductionist materialism that denies human freedom,"[97] his own account provides reasonable grounds for such fears.

By Arnhart's own testimony, Darwin and his acolytes have had an ambivalent record on the issue of reductionism. While Arnhart fails to mention Darwin's belittling of free will as a "delusion," he does cite Darwin questioning why "thought, being a secretion of brain, [is] more wonderful than gravity a property of matter," and he acknowledges the "strong reductionism" advocated by the dean of sociobiology, Harvard's E. O. Wilson.[98] Arnhart even describes emergence as a solution to what he calls "Darwin's problem" of trying to uphold man's unique capacities while insisting they can be completely accounted for through an unbroken chain of "natural causal laws."[99]

But if Darwin had a "problem" avoiding reductionism, and if modern Darwinists like E. O. Wilson advocate "strong reductionism," then perhaps conservative fears of reductionist Darwinism are not illusory after all. Arnhart concedes this point at least implicitly by urging Darwinists to adopt emergence in order to defend human freedom and dignity against reductionism. Yet as long as there is no proof that Darwinists as a whole have followed Arnhart's counsel, why should conservatives relinquish their fears?

Of course, even if Arnhart were to convince Darwinists to repudiate "strong reductionism" in favor of his position of "emergence," it is questionable whether this concession would in fact support a traditional understanding of the human person. If, as Arnhart contends, the emergence of mind required only matter and energy operating according to a purposeless process, then it is difficult to see why the human mind should not be completely reducible in principle to its material building blocks. After all, what else is there in the materialist understanding of nature?

It should be pointed out that the problems posed by Arnhart's concept of emergence reach well beyond the issues of free will and personal responsibility. Emergence as Arnhart describes it also conflicts with the traditional Judeo-Christian belief in an immaterial soul that gives equal dignity to each and every human being, no matter how physically incapacitated. While Arnhart agrees that "only human beings have a soul," he redefines the soul in material terms as simply a large and complex "neocortex, which allows for greater behavioral flexibility."[100] If this is the case, however, what is the worth of human beings with damaged or undeveloped neocortexes? Does a person who has a damaged brain—say, an elderly woman with dementia—merit treatment as a human being? Or can she be treated as a defective dog or cat, which can be euthanized? What about a newborn infant with a genetic defect like Down's Syndrome?

As Arnhart himself notes, Darwinian bioethicist Peter Singer has sought to justify infanticide and euthanasia of mentally defective individuals precisely on the grounds that their damaged brains makes them worth less than lower animals. "Once the religious mumbo-jumbo surrounding the term 'human' has been stripped away," writes Singer, "we will not regard as sacrosanct the life of each and every member of our species, no matter how limited its capacity for intelligent or even conscious life may be."[101] In his view, "[i]f we compare a severely defective human infant with a nonhuman animal, a dog or a pig, for example, we will often find the nonhuman to have superior capacities, both actual and potential, for rationality, self-consciousness, communication, and anything else that can plausibly be considered morally significant."

To his credit, Arnhart tries to refute Singer, but his attempt to do so exposes the ultimate weakness of his own position. Unlike traditional opponents of euthanasia, Arnhart cannot argue for the intrinsic value of handicapped infants or adults. Instead, he condemns Singer for saying that in the case of defective infants "we should put aside feelings based on the small, helpless, and—sometimes—cute appearance of human infants."[102] According to Arnhart, such advice is a mistake because it "assume[s] that we can organize our moral lives around norms derived from abstract reasoning without guidance from our natural emotions."[103] Since our emotions were developed through a long process of evolution, Arnhart believes that ignoring them would be tantamount to going against human nature.

Arnhart's alternative to Singer seems to boil down to "if it feels good, do it"—hardly a position most conservatives would want to embrace. Actually, Arnhart himself recognizes that his argument is insufficient, because he quickly adds: "Of course, when our moral emotions conflict, then we must employ practical reasoning to develop rules of action to resolve the conflict."[104] But this concession eviscerates the force of Arnhart's original objection.

Consider again the case of infanticide. Yes, parents usually have a natural emotional attachment to their baby. But if their baby is seriously defective, they will likely have other emotions as well: They may experience sadness about the child's plight, and pity about his suffering. They may feel anxious and overwhelmed by the burden of dealing with a handicapped child. If the baby's face or limbs are physically deformed, they may even feel revulsion. In other words, those situations where most parents would consider infanticide would be precisely the situations where Arnhart admits that our emotions are insufficient to help us make a moral choice. Given that parents considering infanticide will likely face conflicting emotions, why is Singer's use of "abstract reasoning" inappropriate? Arnhart supplies no convincing answer.

Rightly trained emotions offer important support for moral behavior, and a key purpose of moral training is to habituate people into the proper connections between emotion and moral action. But emotions alone are no substitute for moral reflection and a firm grasp of moral truth.

CHAPTER IV

Does Darwinism Support Economic Liberty?

"Greed… captures the essence of the evolutionary spirit."
Gordon Gekko in the film "Wall Street" (1987).

The idea that Darwinism supports the free enterprise system is deeply embedded in the American imagination, although it is not always described in positive terms. Most people probably first encountered this claim in high school social studies classes, where they heard about ruthless capitalists during the "Gilded Age" who appealed to Darwin's theory of natural selection to justify cut-throat business competition. One of the champions of this philosophy of "Social Darwinism" was William Graham Sumner of Yale, who famously boasted that "millionaires are a product of natural selection," and added that "if we do not like the survival of the fittest, we have only one possible alternative, and that is the survival of the unfittest."[105]

Yet this conventional history of the Gilded Age is more myth than fact. While a few nineteenth century biologists and social theorists justified *laissez faire* economic policies in terms of natural selection, most American defenders of capitalism did not. If anything, they were skeptical about economic applications of Darwin's theory because of its close connection to the Rev. Thomas Malthus's overly pessimistic *Essay on the Principle of Population* (1798).[106]

By Darwin's own account, it was his reading of Malthus that stimulated him to develop his theory of natural selection. Malthus argued that men, animals, and plants all tend to reproduce more offspring than nature can support. The inevitable result of this overpopulation is widespread death until the population is reduced to a level that nature can support. Darwin adopted this struggle for existence articulated by Malthus as the foundation for his theory of evolution by natural selection. Darwin wrote that while reading Malthus, "it at once struck me that under these circumstances [of the struggle for existence] favourable variations would tend to be preserved and unfavourable ones to be be destroyed. The result of this would be the formation of new species."[107]

Applied to the world of commerce, Malthusian theory presented economics as a zero-sum game. Additional people almost inevitably meant greater privation for many human beings. The more people there are, the less food there will be to go around. The more laborers there are, the lower the standard wage will be.[108] While Malthus noted some exceptions to this rule, he suggested they were temporary. In America, for example, "the reward of labour is at present... liberal," but "it may be expected that in the progress of the population of America, the labourers will in time be much less liberally rewarded."[109] In the Malthusian view, economic progress for the few could only be purchased at the price of misery for the many.

American defenders of capitalism during the latter 1800s explicitly repudiated the Malthusian view of economics, which meant that they also had little desire to invoke Darwinism as a defense of free enterprise. In 1879, for example, Harvard political economist Francis Bowen inveighed against "Malthusianism, Darwinism, and Pessimism" in the *North American Review.* Bowen generally supported *laissez faire*, but he was anything but a Malthusian or a Social Darwinist. Contra Malthus, Bowen argued that "the bounties of nature are practically inexhaustible."[110] Therefore starvation and misery among human beings were not inevitable consequences of overpopulation but the products of human ignorance, indolence, and self-indulgence. "It is not the excess of population which causes the misery, but the misery which causes the excess of population," he insisted.[111] Bowen noted that "since 1850... English writers upon political economy have generally ceased to advocate Malthusianism and its subsidiary doctrines," and observed how incongruous it was that "in 1860, at the very time when this gloomy doctrine of 'a battle for life' had nearly died out in political economy... it was revived in biology, and made the basis in that science of a theory still more comprehensive and appalling than that which had been founded upon it by Malthus."[112]

Ironically, it was not capitalism's defenders but its detractors who most vigorously identified capitalism with Darwinian theory. In the late nineteenth and early twentieth centuries, various left-wing reformers tried to discredit capitalism by claiming that it was nothing more than Darwinian "survival of the fittest" applied to the world of business. According to historian Robert Bannister, "[n]ew Liberals and socialists asserted in almost a single voice that opponents of state activity wedded Darwinism to classical economics and thus traded illicitly on the prestige of the new biology."[113] As a result, the primary use of the epithet "Social Darwinism" was not to justify capitalism, but to stigmatize it in order to undermine its legitimacy and generate support for expanded government control over the economy. Darwinism became one of the most potent rhetorical weapons in the arsenal of those who wanted to attack capitalism.

Interestingly, after critics had effectively tarred capitalism with the Social Darwinist label, more businessmen and economists did begin to appropriate the Darwinist metaphor as a defense for free enterprise. By the 1920s, articles in the business press regularly appealed to the Darwinian process as a justification for competition or as a reason against government intervention.[114] At the same time, there was continued resistance to any wholesale appropriation of the Darwinian metaphor among capitalism's defenders.

In his influential work *Socialism* (first published in English in 1932), economist Ludwig von Mises chided the attempt to apply Darwinism and the struggle for existence to economic relations within society. Although men do engage in a struggle against the "natural environment" in order to survive, the purpose of society is to replace that struggle with social cooperation. "Society... in its very conception... abolishes the struggle between human beings and substitutes the mutual aid which provides the essential motive of all members united in an organism. Within the limits of society there is no struggle, only peace."[115] While it is true that the "task" of "economic competition" is "selection of the best," von Mises argued that the Darwinian metaphor was peculiarly inapt as a description of this process, because competition properly understood "is an element of social collaboration" not social warfare.[116] Consequently, von Mises believed it was utterly inappropriate to equate the destruction of uncompetitive businesses with a Darwinian war for survival:

> People say that in the competitive struggle, economic lives are destroyed. This, however, merely means that those who succumb are forced to seek in the structure of the social division of labour a position other than they one they would like to occupy. It does not by any means signify that they are to starve. In the capitalist society there is a place and bread for all. Its ability to expand provides sustenance for every worker. Permanent unemployment is not a feature of free capitalism.[117]

While leading conservatives continued to reject depictions of capitalism as struggle for survival akin what took place in nature, some offered a more sophisticated argument linking Darwinian theory to free enterprise by emphasizing the ability of economic systems to generate "spontaneous order" without an overarching designer. Here Darwin's emphasis on the unguided nature of evolution was regarded as the most relevant application for economics. Accordingly, F. A. Hayek, who wrote dismissively of "Social Darwinism," championed "the emergence of order as the result of adaptive evolution."[118] This was the belief that "complex and orderly and, in a very definite sense, purposive structures might grow up which owed little or nothing to design, which were not invented by a contriving mind but arose from the separate actions of many men who did not know what they were doing."

Arnhart adopts Hayek's idea of "spontaneous order" as a key plank in the platform of "Darwinian conservatism." In Arnhart's view, the concept of "spontaneous order" is not only grounded in the truths of Darwinian biology, but it flatly contradicts the major assumption of "intelligent design": "the fundamental premise of the 'intelligent design' argument is that complex order in the living world must be the deliberately contrived work of an intelligent designer, which denies Hayek's notion of spontaneous order."[119]

Other recent popularizers of a "Darwinian" view of economics also stress the centrality of unguided evolution in business, highlighting in particular the claims of "complexity theory" that complex systems in nature have "self-organizing" properties that can naturally produce even greater levels of complexity.[120] Some libertarians see in complexity theory at least a partial vindication of traditional *laissez faire*.

"On the surface, the computer-assisted discovery of spontaneous order would appear to be a triumphant vindication of libertarian social theory in general and the Austrian School of economics in particular," wrote William Tucker in the libertarian magazine *Reason*.[121] Tucker added that "at the heart of complexity theory... lies the notion of freely evolving systems, including social and economic systems."[122] But Tucker also noted that economists who embrace complexity theory have used it to support government intervention in the economy rather than *laissez faire*. He attributed this to their ideological beliefs, which blinded them to the logical implications of their research.[123]

However, this effort to extrapolate from Darwin's mechanism in nature to the "spontaneous order" found in human society is based on a false analogy. The causes of "spontaneous order" among human beings are simply not equivalent to the mindless process of chance and necessity postulated by Darwinian biology. The Darwinian process in nature is supposed to be blind to intelligence and to the future. Random (i.e., purposeless and non-guided) variations are in the driver's seat. Variations in the social world, however, are driven by human beings exercising their intelligence and foresight. This intelligence and foresight may well be limited, but it is neither purposeless nor completely blind to the future. Moreover, cultural variations simply are not transmitted in a manner equivalent to biological inheritance.

Cultural (as opposed to biological) inheritance depends on learning, teaching, and choice, not on a mechanical process of genetic transmission. If anything, so-called Darwinian analogies applied to business could be better described as Lamarckian analogies, because they involve the transmission of characteristics acquired through an organism's conscious efforts to adapt to its environment. While it is true that social cooperation may not be guided by a single designer, that is not because the process is driven by random variations but because it results from the intelligent choices of innumerable designers interacting with each other.

It is somewhat misleading to say that human order arises "spontaneously." That term seems to suggest a lack of any conscious thought, yet human social order arises out of the intentional actions of individuals and groups to associate with each other, to exchange goods, and to improve their environment. Just because these actions are decentralized does not mean they are not designed. This sort of design-driven cooperation is alien to the Darwinian mechanism. It is not alien, however, to the teachings of the classical school of economics that predated Darwin.

While Arnhart asserts that "[a] spontaneous order is an unintended order" and that "[s]pontaneous order is design without a designer,"[124] he implicitly acknowledges that this is not literally true. At one point he writes about "[a]llowing social order to arise spontaneously through the *mutual adjustment of individuals and groups seeking their particular ends.*"[125] But order that arises through the "mutual adjustment of individuals and groups" who are pursuing "their particular ends" does not come about through a purposeless interaction of chance and necessity. It comes about through the rational actions of many intelligent designers.

Arnhart's own examples of "spontaneous order" merely underscore this point. For example, he holds up the evolution of the English language as "spontaneous order." Yet what he actually means by this is that "[o]ur language has been enriched by a few great minds like William Shakespeare... and by the many small minds of ordinary people in ordinary speech."[126] Arnhart is right, but in a way he apparently does not realize. The English language has developed through the interaction of many *minds*—not through a roll of the dice, not through a group of monkeys typing away at a 100 computer keyboards, and certainly not through a blind process of natural selection acting on random mutations. The fact that English was not developed by "a group of English linguists who could reform our language from the top down,"[127] does not mean it was the result of chance and necessity rather than design.

Arnhart here seems to conflate the lack of an *overarching* design with the absence of any design. But something can still be the product of intelligent causes even if it is not the product of a single omnicompetent designer. Arnhart's description of the evolution of English is a good example, as is the web phenomenon known as "Wikipedia." Wikipedia's content is not the product of one overarching designer, but the work of many rational minds operating in collaboration. Wikipedia is not a demonstration of the power of the Darwinian mechanism of chance and necessity; it is a demonstration of the power of intelligent causes working together.

Contrary to Arnhart, the idea of intelligent design is perfectly compatible with the notion of "spontaneous order" arising in society from the actions of multiple rational agents operating within a context of limited knowledge and power.[128] It is Darwinism's unguided process of selection and mutation that poses the real problem for "spontaneous order," because it asserts that complex order can arise without any goal-directed actions at all, thus discounting the need for the purposeful interaction of rational agents on which the spontaneous order found in the human world depends.

CHAPTER V

DOES DARWINISM SUPPORT "NON-UTOPIAN LIMITED GOVERNMENT"?

"[A]ll that progressives ask or desire is... to interpret the Constitution according to the Darwinian principle."
WOODROW WILSON, *THE NEW FREEDOM*.[129]

One of the most extraordinary claims made by Darwinian conservatives is their insistence that Darwinian theory supports realism and limited government rather than utopian efforts to transform society. In truth, scientists and planners during the past century have drawn on Darwinian theory to promote one utopian crusade after another, including forced sterilization, scientific racism, euthanasia, and an ever-expanding government justified in the name of the "evolving Constitution."[130]

The typical response of Darwinists to this record of coercive "Social Darwinism" is to deny that it has any genuine connection to Darwin or his theory of evolution. This is Arnhart's basic position: "We need to recognize... that much of what has been identified as social Darwinism or the Darwinian left is a distortion of Darwinian science."[131] But when one examines the historical record in detail, the effort to disentangle Darwinism from "Social Darwinism" is hard to maintain. This can be be be seen most clearly in the case of eugenics.[132]

Eugenics and Darwinism.

Eugenics was promoted as the science of human breeding, and during the first several decades of the twentieth century, it resulted in the compulsory sterilization of more than 60,000 presumed "defectives" in the United States, including many who probably would not be considered mentally deficient today.[133] Racial minorities and the poor were special targets of the eugenics crusaders, and their program of forced sterilization was ultimately approved by the U. S. Supreme Court in the infamous case of *Buck v. Bell*, where Supreme Court Justice Oliver Wendell Holmes Jr. declared that "three generations of imbeciles are enough."[134] Eugenists attempted to solve the nation's social ills by scientifically breeding better human beings. By the time the eugenics crusade ended, it had left a trail of broken lives, junk science, and dehumanizing rhetoric whose legacy is still felt today.

Charles Darwin's cousin Francis Galton is generally considered the founder of the modern eugenics movement. After researching the family connections of members of the British elite, Galton announced that talent was largely hereditary.[135] Thus, if society wanted to guarantee its future improvement, it needed to pay attention to those who were having the most babies. By the 1880s Galton had coined the actual term eugenics (adapted from a Greek root word meaning "good in birth"[136]), and he was urging efforts to improve the race through better breeding. "Positive eugenics" focused on encouraging those deemed the most fit to reproduce more, while "negative eugenics" focused on curtailing reproduction by those deemed unfit, including mental defectives and criminals.

The eugenics movement drew direct inspiration from Darwinian biology, and it was promoted first and foremost by Darwinian biologists. Yet today the Darwinian roots of eugenics tend to be downplayed both by the popular media and by the defenders of Darwinism. When Darwin's theory is mentioned at all, a sharp distinction is often drawn between Darwin's own views and the "Social Darwinism" of the eugenists, who supposedly extended Darwin's theory into realms unanticipated or even opposed by Darwin.

In the recent book *War Against the Weak*, for example, Edwin Black argues that "Darwin was writing about a 'natural world' distinct from man," while other thinkers were the ones to blame for "distilling the ideas of Malthus, Spencer and Darwin into a new concept, bearing a name never used by Darwin himself: *social Darwinism*."[137] Black seemed unaware that Darwin wrote extensively about the application of natural selection to human beings in *The Descent of Man*. But at least Black acknowledged the influence of Darwinism. Sometimes today the connection between Darwinian biology and eugenics is evaded altogether.

On the "Understanding Evolution" website funded by the National Science Foundation, users will find a cartoon showing Charles Darwin yelling "Get out of my house!" to a proponent of eugenics.[138] The point is clear: Darwin opposed eugenics. Incredibly, one educator writing recently about eugenics not only failed to mention Darwinian biology, he traced the eugenists' beliefs instead back to the Bible! In his view eugenics embodied "the biblical concept that 'like breeds like,' to which eugenics researchers provided a scientific gloss."[139]

Yet it was society's violation of natural selection, not the Bible, that supplied the operating premise for the eugenists' ideology. The eugenists' underlying fear was the same as the one Charles Darwin himself had articulated in the following passage in *The Descent of Man*:

> With savages, the weak in body or mind are soon eliminated... We civilised men, on the other hand, do our utmost to check the process of elimination; we build asylums for the imbecile, the maimed, and the sick; we institute poor-laws; and our medical men exert their utmost skill to save the life of every one to the last moment. There is reason to believe that vaccination has preserved thousands, who from a weak constitution would formerly have succumbed to small-pox. Thus the weak members of civilised societies propagate their kind. No one who has attended to the breeding of domestic

> animals will doubt that this must be highly injurious to the race of man... hardly any one is so ignorant as to allow his worst animals to breed.[140]

Arnhart downplays the importance of this startling passage by noting that Darwin added that we can't simply follow these dictates of "hard reason" without destroying the "noblest part of our nature," our sympathy.[141]

But this was a rather lame objection on Darwin's part. If Darwin believed that society's efforts to help the impoverished and sickly "must be highly injurious to the race of man" (note the word "must"), then the price of preserving compassion in his view appeared to be the destruction of the human race. Framed in that manner, how many people could be expected to reject the teachings of "hard reason" and sacrifice the human race?

For Darwin himself, the objection from compassion did not appear to be decisive. He spent the rest of his discussion of natural selection and modern society trying to argue that natural selection still kills off enough defective individuals to prevent racial degeneration: "Malefactors are executed, or imprisoned for long periods, so that they cannot freely transmit their bad qualities. Melancholic and insane persons are confined, or commit suicide. Violent and quarrelsome men often come to a bloody end... Profligate women bear few children, and profligate men rarely marry; both suffer from disease."[142] Darwin was at his least convincing when he tried to assuage fears by claiming that the profligate married and reproduced less than those he considered healthy and virtuous. Later Darwin himself indicated otherwise. "The reckless, degraded, and often vicious members of society, tend to increase at a quicker rate than the provident and generally virtuous members."[143] Darwin nevertheless hoped that the higher mortality rate of such classes might prevent them from overwhelming "the better class of men." But he was not so bold as to promise such a result. Instead, he stated that if natural checks did not prevent the inordinate growth of degraded classes, "the nation will retrograde, as has occurred too often in the history of the world. We must remember that progress is no invariable rule.... Natural selection acts only in a tentative manner."[144]

Perhaps fearing that this comment was too grim for Victorian sensibilities, in his revised edition of *The Descent of Man* Darwin tacked on a happy conclusion to the discussion, now asserting that "the more intelligent members within the same community will succeed better in the long run than the inferior, and leave a more numerous progeny, and this is a form of natural selection."[145] That statement verged on the disingenuous. Not only did it cut against much of the evidence Darwin marshalled, but it contradicted Darwin's own private views during the latter part of his life. According to Alfred Wallace, during one of his last conversations with his friend, Darwin

> expressed himself very gloomily on the future of humanity, on the ground that in our modern civilisation natural selection had no play, and the fittest did not survive. Those who succeed in the race for wealth are by no means the best or the most intelligent, and it is notorious that our population is more largely renewed in each generation from the lower than from the middle and upper classes.[146]

It is possible that Darwin became more negative on this subject after his revised edition of *Descent of Man* was published, and this increased negativism was reflected in Wallace's account; but it seems more likely that Darwin harbored the same doubts while he wrote *Descent*. He simply repressed them when writing for the public.

It would be anachronistic to regard Darwin as a full-blown eugenist. Surgical sterilization, which became the primary tool of negative eugenics, did not become technically feasible until after Darwin died. It is also true, as Arnhart points out, that Darwin believed that positive eugenics was somewhat utopian, although he praised the work of his cousin Francis Galton and urged further scientific research that might make his goals less utopian.[147]

Nevertheless, it is clear that Darwin laid the groundwork for at least negative eugenics in *The Descent of Man*. Darwin emphasized that man had achieved his current rank in the animal world through natural selection,[148] and he further stressed that the human race would degenerate if the pressure of natural selection was relieved:

> Man, like every other animal, has no doubt advanced to his present high condition through a struggle for existence consequent on his rapid multiplication; and if he is to advance still higher he must remain subject to a severe struggle. Otherwise he would soon sink into indolence, and the more highly-gifted men would not be more successful in the battle of life than the less gifted.[149]

Darwin did maintain (at least in public) that man had not yet so counteracted natural selection that the destruction of the human race was imminent. But he worried that if civilized societies continued down the path of preserving the weak a crisis loomed, and in private he fretted that the crisis had already arrived.

By the early 1900s, eugenists believed they had strong evidence that the crisis foreseen by Darwin had come to pass. In their view, social welfare efforts had developed sufficiently to make Darwin's fears a reality. Time and again, they lamented civilized society's increasing sins against natural selection.

According to former Governor of Illinois Frank Lowden, "in a state of nature" defective individuals "would long ago have disappeared from the face of the earth. Starvation, disease, and exposure, if they had been left to their own resources, would have eliminated them long ago. Man's interference with natural laws alone save them from perishing."[150]

Edwin Conklin, Professor of Biology at Princeton University, observed that while nature may still kill off the worst defectives, "nevertheless a good many defectives survive in modern society and are capable of reproduction who would have perished in more primitive society before reaching maturity."[151] Such defectives survive "in the most highly civilized States" because they "are preserved by charity, and... are allowed to reproduce... thus natural selection, the great law of evolution and progress, is set at naught."

Some eugenists even invoked natural selection to criticize efforts to reduce infant mortality by improving sanitation, hygiene, and prenatal care. According to these critics, such efforts merely postponed the deaths of many defective babies, and those defective babies who did survive long-term would drag the race down by perpetuating "another strain of weak heredity, which natural selection would have cut off ruthlessly in the interests of race betterment."[152] Hence, "from a strict biological viewpoint" efforts to reduce infant mortality by improving environmental influences were "often detrimental to the future of the race." Prof. H. E. Jones of the University of Virginia made the same point more generally: "What sanitary science and hygiene seek to accomplish by attention to external conditions alone largely defeats its own ends by counteracting the working of the principle of selection."[153]

Eugenists believed that social welfare programs were helping defectives and the lower classes to breed at faster rates, and soon they would swamp the rest of society. Given this situation, eugenists thought society had only two choices: Go back to the law of the jungle and stop preserving the weak; or institute some form of artificial selection (eugenics) to replace the absence of natural selection. Both options were grounded in a thoroughly Darwinian understanding of human society. Eugenics may not have been the only possible application of Darwin's theory; but it certainly was logically connected to Darwin's theory. Far from a distortion of Darwin's ideas, it was a culmination of the warnings issued by Darwin himself in *The Descent of Man,* and that is how eugenists understood it, especially in the United States.

The leaders of the eugenics crusade were largely university-trained biologists and doctors, not politicians, and they pushed for eugenics because they thought it was fully justified by Darwinian biology. "Eugenics is a branch of biology—social biology—and its study has been cultivated chiefly by the biologists," insisted biologist Charles Davenport.[154] "The biologist... demands cures instead of first-aid," added Harvard biologist Edward East, who condemned most social service programs as "unsound biologically" and justified eugenic birth-control as the appropriate scientific alternative.[155]

It should be stressed that leading eugenists represented mainstream evolutionary biology, not the fringe. They were affiliated with institutions like Harvard, Princeton, Columbia, and Stanford.[156] They were leaders in America's most prestigious scientific organizations. Biologist Edwin Conklin was president of the American Association for the Advancement of Science (AAAS). Paleontologist Henry Fairfield Osborn was director of the American Museum of Natural History in New York, which sponsored an extensive exhibit promoting eugenics during the Third International Congress of Eugenics. In sum, eugenists were members of the scientific establishment, and their views became so dominant for a time that eugenics was for all practical purposes the "consensus" view of the scientific community. Accordingly, lectures and courses on eugenics were taught at many American colleges and universities, and the topic was included as a standard part of high school and college biology textbooks.[157]

Those who insist that eugenics was somehow a distortion of Darwinian biology must account for the fact that the vast majority of leading Darwinian biologists for several decades thought otherwise. Indeed, they promoted the tenets of eugenics as "strict corollaries" of "the theory of organic evolution."[158] Were these Darwinian biologists all wrong? If so, why? If Darwinian biology really was hostile to eugenics, why weren't there many prominent evolutionary biologists who said so?

It is difficult to find any prominent Darwinian biologist from the first part of the twentieth century who fundamentally opposed eugenics. Some expressed doubts about positive eugenics and the ability to breed a superior race, but few if any opposed negative eugenics such as forced sterilization. Princeton's Edwin Conklin is a good example. Conklin was skeptical that eugenics could usher in a utopia, but he continued to attack modern society for encouraging "the perpetuation of the worst lines through sentimental regard for personal rights, even when opposed to the welfare of society," and he urged society "to substitute intelligent artificial selection for natural selection," since natural selection was "in so far as possible nullified by civilized man."[159] Whatever his doubts about the ultimate efficacy of eugenics, Conklin continued to embrace the movement's major public policy goal, which was the prevention of propagation by the unfit.[160]

However distasteful it may be to modern Darwinists, the eugenics crusade—especially negative eugenics—was a logical application of the teachings of Darwinian biology, and no amount of rationalizing will get around that truth.

But eugenics was not the only utopian crusade with ties to Darwinian biology. As Arnhart himself notes, "[m]any of the people who were prominent in the history of social Darwinism and eugenics were socialists. So there seems to have been a 'Darwinian left.'"[161]

Socialist Darwinism.

Many on the far left were attracted to Darwinian theory first of all because of its seeming confirmation of a materialistic understanding of the natural world. According to Frederick Engels, "Darwin first developed in connected form" the "proof... that the stock of organic products of nature environing us today, including man, is the result of a long process of evolution from a few originally unicellular germs, and that these again have arisen from protoplasm or albumen, which came into existence by chemical means."[162] Darwin's materialistic theory was also praised by socialists for banishing purpose from nature. According to Marx, "[d]espite all shortcomings, it is here [in Darwin's work] that, for the first time, 'teleology' in natural science is not only dealt a mortal blow but its rational meaning is empirically explained."[163] Engels agreed.

Before Darwin, he wrote, "[t]here was one aspect of teleology that had yet to be demolished, and that has now been done."[164]

In addition to providing a convincing materialistic account of the natural world, some on the left thought that Darwin's theory of natural selection provided a biological foundation for the class struggle in human society. "Although it is developed in the crude English fashion," *The Origin of Species* "is the book which, in the field of natural history, provides the basis for our views," Karl Marx wrote Frederick Engels in 1860.[165] "Darwin's work is most important and suits my purpose in that it provides a basis in natural science for the historical class struggle," he added to another correspondent in 1861.[166] While Marx ultimately expressed ambivalence about the relevance of Darwin's theory to human affairs, he was not above invoking natural selection to help explain specific economic struggles in *Das Kapital* (1867).[167] Other socialists drew similar connections.[168]

In the socialist version of survival of the fittest, it was the members of the proletariat who would prove themselves fittest by joining together to topple their capitalist overlords. In the words of Engels, "[t]he struggle for existence can then consist only in this: that the producing class takes over the management of production and distribution from the class that was hitherto entrusted with it but has now become incompetent to handle it, and there you have the socialist revolution."[169]

Reinforcing the class struggle with Darwinism was not universal among those on the left,[170] and even those who made the connection were not arguing for permanent social conflict of the kind found in the natural world. As Gertrude Himmelfarb points out, "the struggle for existence that Darwin took to be a permanent condition of animal life, Marx saw as a condition only of particular epochs in human history," and he assumed that this struggle "would be suspended or transcended to permit the emergence of the classless society."[171] So while Darwin's theory might help explain some of the ultimate roots of class conflict, it did not provide a prescription for social policy. The point of socialism and Marxism was to change the world, not merely accept the status quo.[172] For socialists and Marxists, the materialistic understanding of history did not lead to a bleak endorsement of current conditions, but to scientific knowledge about how to reshape the material world to fulfill mankind's boldest dreams.

Arnhart argues that socialist efforts to appropriate Darwinism all failed "because of the irreconcilable conflict between Darwinian science and socialist utopianism." He insists that "[t]he socialist belief in human perfectibility must deny a Darwinian science of human nature that constrains the human freedom for utopian transformation." But is it really clear that a Darwinian view of human nature is incompatible with schemes for "utopian transformation"? After all, Darwinian evolution is driven as much by fast-changing environmental imperatives as by slow-changing heredity. Current population characteristics are adaptive only insofar as the environment stays the same. Whenever the environment changes, there is the possibility that new traits will be required for survival. Once one understands this truth, why can't one hope to reorder the environmental conditions of human society in such a way as to change which human characteristics are most desirable?

Indeed, by revealing to us the truth by which new human traits come into being, Darwinian biology would seem to offer encouragement to those who want to transform society. Now that they know how new traits actually develop, they can seek to harness the evolutionary process for their own ends. Socialist Darwinism may not be as closely connected to Darwinian biology as eugenics, but socialist interpretations of evolution do not seem necessarily incompatible with Darwinism. The same is true of the use of Darwin by mainstream liberals, especially the progressive movement, in the United States.

Progressivism and the Darwinian Constitution.

What American progressives found most useful about Darwin's theory was the case it made for the necessity—indeed, the inevitability—

of change. According to the progressives, nations and their economic systems are subject to evolution just as much as plants and animals; and like plants and animals, nations must adapt to new conditions or die. Governments must evolve in order to deal with new challenges.

The roots of the progressive idea of social and political evolution were supplied not by Darwin but by Hegel and the political science of the German administrative state.[173] But Darwin was honored for showing that the truths preached by the political philosophers had been substantiated by biology. This faith in the necessity of social evolution underlay the progressive movement in America and its rejection of *laissez faire* capitalism. While *laissez faire* may have been required by the exigencies of a previous era, new social conditions called for new economic policies according to the progressives, and therefore American government must evolve and grow in order to meet the demands of the new circumstances.

In adopting their evolutionary view of government, the progressive reformers had to discard the political theory of the American founders just as much as they did the teachings of *laissez faire* capitalism. The founders had claimed that government existed to secure certain unchanging natural rights. Because these rights did not change, neither did the purposes of the government. The founders' conception of government was thus more stationary than evolutionary. But such a conception was tantamount to heresy in the new scientific view of the world.[174]

One of the most articulate spokesmen for the new view was a political scientist from New Jersey, who argued that "in our own day, whenever we discuss the structure or development of a thing... we consciously or unconsciously follow Mr. Darwin."[175] The political scientist was Woodrow Wilson, then president of Princeton, soon to be Governor of New Jersey and eventually President of the United States. During the presidential election campaign of 1912, Wilson explicitly invoked Darwin to justify an evolutionary understanding of the U.S. Constitution that would allow the federal government to dramatically expand its powers over the economy.

According to Wilson, the problem with the original Constitution was that it betrayed the founders' "Newtonian" view that government was built on unchanging laws like "the law of gravitation."[176] In truth, however, government "falls, not under the theory of the universe, but under the theory of organic life. It is accountable to Darwin, not to Newton. It is modified by its environment, necessitated by its tasks, shaped to its functions by the sheer pressure of life." Hence, "living political constitutions must be Darwinian in structure and in practice. Society is a living organism and must obey the laws of Life... it must develop." According to Wilson, "all that progressives ask or desire is permission—in an era when 'development,' 'evolution,' is the scientific word—to interpret the Constitution according to the Darwinian principle." The doctrine of the evolving Constitution articulated by Wilson and other progressives opened the door to much greater regulation of business and the economy, eventually paving the way for the New Deal.

CHAPTER VI

Is Darwinism Compatible with Religion?

"Darwin made it possible to be an intellectually fulfilled atheist."
RICHARD DAWKINS, *THE BLIND WATCHMAKER*.[177]

Religious believers are a core constituency of the conservative movement, and most conservatives, religious or not, uphold the social importance of religion as a support for morality and free government. Hence, if Darwinism is to be used to support conservatism, any presumed hostility toward religion must be eliminated. Arnhart argues accordingly that "Darwinian biology is compatible with religious belief, and particularly with biblical theism."[178] James Q. Wilson agrees: "Evolution rules out the possibility that God created each species one at a time but it does not rule out the possibility that God designed natural selection, infused mankind with a soul, or presides over an afterlife."[179] Charles Krauthammer, meanwhile, faults the critics of Darwin for unfairly suggesting that evolution and religion conflict: "How ridiculous [it is] to make evolution the enemy of God."[180]

Darwin's conservatives are not alone in attempting to square the theory of evolution with religion. Recognizing that most Americans believe in God, and that if evolution is seen as anti-religious it will continue to lag in popular support, the National Center for Science Education has spearheaded a campaign to convince religious believers that evolution and religion are compatible. On a federally-funded website that the NCSE helped design, teachers and students are directed to a list of statements by religious groups endorsing evolution, and Eugenie Scott, the group's executive director, encourages biology teachers to spend class time having students read statements by religious leaders supporting evolution. Scott even suggests that students be assigned to interview local ministers about their views on evolution—but not if the community is "conservative Christian," because then the lesson that "Evolution is OK!" may not come through.[181] The NCSE's effort to inject religion into public school science classes in order to promote evolution is a remarkable act of chutzpah for an organization that routinely chastises "antievolutionists" for supposedly trying to insert "religion" into science classes. Apparently, religion in biology class is OK so long as it is used to endorse Darwin's theory.

The NCSE also advises inviting ministers to testify before school boards in favor of evolution,[182] and it has created a curriculum to promote evolution in the churches.[183] The NCSE even has a "Faith Network Director" who claims that "Darwin's theory of evolution... has, for those open to the possibilities, expanded our notions of God."[184] Other evolutionists have collected signatures from liberal clergy in support of evolution as part of "The Clergy Letter Project" and have urged churches to celebrate "Evolution Sunday" on the Sunday closest to Darwin's birthday.[185]

This attempt to put a religious face on modern evolutionary theory is an effort to deal with what might be called Darwinism's "Dawkins' problem." Oxford biologist Richard Dawkins is one of the world's foremost boosters of Darwinian evolution. Unfortunately for evolutionists, Dawkins zealously expounds the anti-religious implications of the theory, and regularly denounces religion. One of his choicer comments is his description of religious faith as "one of the world's great evils, comparable to the smallpox virus but harder to eradicate."[186] By highlighting the religious defenders of evolution, the NCSE undoubtedly hopes to depict Dawkins as a fringe figure whose views are not representative of Darwinists as a whole.

A key problem with this depiction is that Dawkins is far from unrepresentative of the views of prominent Darwinists. The public relations efforts of the NCSE notwithstanding, a dominant majority of leading defenders of Darwinism seem to be either avowed atheists or agnostics. Barbara Forrest, co-author of *Creationism's Trojan Horse: The Wedge of Intelligent Design,* is a long-time board member of the New Orleans Secular Humanist Association, which describes itself as "an affiliate of American Atheists, and [a] member of the Atheist Alliance International."[187] Physicist Victor Stenger, author of *Not by Design: The Origin of the Universe,* urges his fellow scientists "to make a strong, scientific statement about the very likely nonexistence of the Judeo-Christian-Islamic God."[188] Geologist Steven Schafersman, head of the pro-Darwin group "Texas Citizens for Science," describes himself as a "secular humanist"[189] and maintains that "Supernaturalistic religion and naturalistic science... are and will remain in eternal conflict."[190] Nobel laureate Steven Weinberg, who championed Darwinism before the Texas State Board of Education in 2003, believes that the downfall of religion is probably "the most important contribution" science can make to the world. In his own words, "I personally feel that the teaching of modern science is corrosive of religious belief, and I'm all for that! One of the things that in fact has driven me in my life, is the feeling that this is one of the great social functions of science—to free people from superstition."[191] Lest there be any doubt about what Weinberg means by "superstition," he goes on to say that he hopes "that this progression of priests and ministers and rabbis and ulamas and imams and bonzes and bodhisattvas will come to an end, that we'll see no more of them. I hope that this is something to which science can contribute and if it is, then I think it may be the most important contribution that we can make."

Even Eugenie Scott, who now tries to convince the public that evolution and religion are harmonious, is a signer (along with Richard Dawkins!) of a document called the "Humanist Manifesto III," which celebrates "the inevitability and finality of death" and proclaims that "humans are... the result of unguided evolutionary change."[192] By specifically citing "unguided evolutionary change" as part of its case for "a progressive philosophy of life... without supernaturalism," this manifesto clearly suggests that evolution properly understood contradicts belief in a personal God.

Survey research of the nation's leading scientists seems to corroborate the anti-religious attitude prevalent among biologists. According to a poll of scientists listed in *American Men and Women of Science*, 57.5% of the biologists who responded were atheists or agnostics and 59.4% disbelieved or were agnostic about personal immortality. The nation's most elite biologists are even more atheistic. According to a 1998 survey of members of the National Academy of Sciences (NAS), 94.4% of the NAS biologists are atheists or agnostics. A similar percentage rejects life after death.[193] By contrast, the vast majority of Americans continue to believe both in God and in personal immortality.[194]

If leading Darwinists tend to be anti-religious today, so too do the grassroots activists. This fact can be seen by the list of groups sponsoring annual "Darwin Day" celebrations to mark the birthday of Charles Darwin each February. The list is top-heavy with organizations bearing such names as the "San Francisco Atheists," the "Gay and Lesbian Atheists and Humanists," the "Humanists of Idaho," the "Central Iowa Skeptics," the "Southeast Michigan Chapter of Freedom from Religion Foundation," the "Long Island Secular Humanists," and the "Atheists and Agnostics of Wisconsin."[195]

Of course, many Christians and other religious believers have embraced "evolution" too. Don't these "theistic evolutionists" prove the compatibility of Darwin's theory and traditional religion? Not really. On closer inspection such religious believers either reject full-blown "Darwinian" (i.e., unguided) evolution or they jettison traditional theism in order to uphold a consistent Darwinism. Consider the case of astronomer and Catholic priest George Coyne, the former director of the Vatican Observatory. [196] A strong defender of Darwin's theory, Coyne is often cited in the newsmedia to show (wrongly) that the Catholic Church endorses Darwinian evolution. But in order defend a truly unguided evolution, Coyne appears to deny traditional Christian doctrines of God's omnipotence and omniscience:

> If we take the results of modem science seriously, it is difficult to believe that God is omnipotent and omniscient in the sense of the scholastic philosophers... Let us suppose that God possessed the theory of everything, knew all the laws of physics, all the fundamental forces. Even then could God know with certainty that human life would come to be? If we truly accept the scientific view that, in addition to necessary processes and the immense opportunities offered by the universe, there are also chance processes, then it would appear that not even God could know the outcome with certainty.[197]

So in Coyne's view, God could not even know beforehand that human beings would be produced by the evolutionary process.

Father Coyne shows how difficult it can be for theistic evolutionists who take Darwin seriously to maintain their traditional religious commitments. Sometimes they completely give up trying to do so. That is what happened to retired Calvin College professor Howard Van Till, who Arnhart cites as prime example of the "Christian evolutionists" who embrace Darwin. Arnhart apparently does not realize that Van Till's beliefs have now evolved well beyond traditional Christianity. According to a lecture he recently delivered to the Freethought Association of West Michigan, he now considers himself a freethinker.[198]

Those who argue that Darwinian theory is compatible with religion have to account for far more than Richard Dawkins. They need to explain why the dominant majority of leading proponents of Darwinism seem to combine it with atheism or agnosticism. Perhaps these Darwinists reject religion on grounds completely unrelated to Darwinism, or perhaps they are all guilty of sloppy logic. But the association between Darwinists and the rejection of religion at least raises a serious question about the presumed harmony between Darwinian evolution and religion.

Other than dismissing Richard Dawkins for offering "almost no evidence to support his passionate assertions that science and religion conflict," Arnhart offers little explanation for why the ranks of Darwinists are so dominated by those who oppose traditional religion.[199] He does try to make the case that conservative Christians in the past had no difficulty reconciling their beliefs with Darwinism: "In fact, throughout the second half of the nineteenth century and the early decades of the twentieth century, many conservative Christians in Britain and America accepted Darwin's teaching as compatible with orthodox Christianity, and thus they adopted various conceptions of theistic evolution."[200]

But this claim is misleading. While many orthodox Christians during the period in question accepted Darwin's idea of descent with modification, they rejected Darwin's mechanism of unguided natural selection acting on random variations as an adequate explanation for the complexity of life. As historian Peter Bowler points out, what made it possible for many religious believers to accept "evolution" during the initial decades after Darwin "was the belief that evolution was an essentially purposeful process... The human mind and moral values were seen as the intended outcome of a process that was built into the very fabric of nature and that could thus be interpreted as the Creator's plan."[201]

Interestingly, even Asa Gray, who is often regarded as the most important "theistic evolutionist" in America to support Darwin, was in fact skeptical of the ability of natural selection and random variation to produce complex organs like the eye. According to historian Ronald Numbers, "Gray confessed to a friend that this theistic version of evolution was '*very anti-Darwin*'."[202]

In other words, many traditional religious believers were able to find common ground with "evolution" precisely because they rejected Darwin's unguided mechanism and embraced a teleological form of evolution. This was possible because many scientists of the time also remained deeply skeptical about the extent to which natural selection and random variation could explain the development of fundamentally new biological features. What brought this era of accommodation to a close was the resurgence of Darwinian natural selection in the early 1900s, fueled by work in experimental genetics.[203]

So just how compatible is modern evolutionary theory with faith in God?

In addressing this question, one needs to return again to the issue of definitions. If by "evolution" one means biological common descent, then surely evolution is compatible with most forms of theism, although perhaps not with a completely literal reading of the book of Genesis. If "evolution" means that natural selection can produce many small changes in existing species, there is even less of a problem. Not even Biblical creationists deny that such "microevolution" can occur. But if by "evolution" one means that all life was developed through an unguided process of chance and necessity, with no particular end in view, then it seems much more difficult to square evolutionary theory with religion, at least in its Judeo-Christian form.

Of course, there are still some possible ways to reconcile robust Darwinism with religious faith. The first option is to insist that evolution is indeed guided by God, but that His guidance is hidden from us. In other words, while the development of life may appear to be the product of chance and necessity, it is in fact following a plan that we cannot detect. This is the view promoted by Francis Collins, head of the Human Genome Project, in his recent book *The Language of God*. "[E]volution could appear to us to be driven by chance," writes Collins, "but from God's perspective the outcome would be entirely specified."[204]

While Collins' view is logically compatible with the idea that God actively guides the development of His creation, it is still in tension with the traditional Biblical understanding of God. Both the Old and New Testaments teach that human beings can recognize God's handiwork in nature through their own observations rather than special divine revelation. "The heavens declare the glory of God; And the firmament shows His handiwork," proclaimed the psalmist.[205] The apostle Paul likewise argued that "since the creation of the world His invisible attributes are clearly seen, being understood by the things that are made."[206] The idea that God's action in the world is in principle undetectable by us seems hard to reconcile with the traditional Judeo-Christian view that God's design in nature is clearly evident to all human beings through the use of their reason.

There is an equally serious difficulty for the idea of undetectable design from standpoint of evolutionists: Darwinism proclaims that evolution is blind and unguided. Postulating that evolution is guided but undetectable essentially guts this claim. Furthermore, if evolution truly is guided, it is hard to see how one can maintain that such guidance is in principle undetectable and will remain so forever. If evolution is guided, how do we know that it cannot be detected some day? It is unsurprising, then, that committed Darwinists who espouse religious beliefs are loathe to adopt this position.

Roman Catholic biologist Kenneth Miller considers the position of guided but undetectable design in his book *Finding Darwin's God*, but rejects it. "Evolution is a natural process, and natural processes are undirected," he insists.[207] Miller denies that the evolutionary process was directed in order to produce any particular result—even the development of human beings. In fact, he says he agrees with the view "that mankind's appearance on this planet was *not* preordained, that we are here not as the products of an inevitable procession of evolutionary success, but as an afterthought, a minor detail, a happenstance in a history that might just as well have left us out."[208]

There is, however, another way to try to resolve the tension between Darwinism and religion. Darwinian evolution, strictly speaking, begins after the first life has developed, and so I agree with Arnhart that it does not necessarily refute the claim that there may some kind of "first cause" to the universe that stands outside of "nature." But this "first cause" allowable by Darwinism seems incompatible with the God of the Bible. It cannot be a God who actively supervises or directs the development of life. The most it could do is to set up the interplay between chance and necessity, and then watch to see what the interplay produces. Such an absentee God is hard to reconcile with any traditional Judeo-Christian conception of a God who actively directs and cares for His creation. In the end, the effort to reconcile Darwinism with traditional Judeo-Christian theism remains unpersuasive.

CHAPTER VII

Has Darwinism Refuted the Challenge from Intelligent Design (ID)?

"[W]e find that ID is not science and cannot be adjudged a valid, accepted scientific theory... ID, as noted, is grounded in theology, not science."
Judge John E. Jones, *Kitzmiller v. Dover* (2005).[209]

Regardless of Darwinism's political, social, or even theological implications, the primary question is still whether it is true. In recent decades, a growing number of scientists and philosophers of science have voiced skepticism of key parts of neo-Darwinism, including its central claim that natural selection and random mutation are sufficient to explain the intricate and highly-functional complexity we see throughout the natural world. Some of these scholars are proponents of what is known as "intelligent design," which proposes that "certain features of the universe and of living things are best explained by an intelligent cause, not an undirected process such as natural selection."[210]

Arnhart rejects "the scientific argument against Darwinism" made by intelligent design proponents because he "believe[s] that the evidence and arguments for Darwin's theory sustain it as a probable truth, although it cannot be conclusively demonstrated."[211] Nevertheless, he accepts what he calls the "political argument" offered by ID proponents that "teaching intelligent in the public schools... would promote the sort of freedom of thought required in a democratic society."

Here it should be noted that the leading national organization supportive of intelligent design, Discovery Institute, opposes requiring ID to be taught in public schools, and has done so for some time. While the Institute believes that objectively discussing the debate over design should be constitutionally permissible, it does not advocate forcing intelligent design into classrooms. Instead, it favors the more limited policy of requiring mainstream scientific criticisms of neo-Darwinism to be presented alongside the best evidence favoring Darwin's theory.[212] Since these criticisms of neo-Darwinism are already well-attested in the standard scientific literature, it is hard to see why students should be prevented from learning about them.

Discovery Institute articulated this limited "teach the controversy" approach to evolution years before the federal court decision in *Kitzmiller v. Dover.* Indeed, the Institute's opposition to mandating ID was a key reason it publicly opposed the Dover school board policy on intelligent design and urged its repeal.[213]

Despite the opposition of leading ID proponents to mandating intelligent design, Arnhart is on target when he observes that "[a] lively classroom debate over Darwinism... might actually prepare students to become citizens capable of judging scientific disputes that have deep consequences for human life."[214] Arnhart shows more robust commitment to intellectual freedom on this issue than many of the self-proclaimed civil libertarians at groups like the American Civil Liberties Union and Americans United for the Separation of Church and State.

Arnhart also helps expose the fact that in the current debate it increasingly is the Darwinists who want to limit the teaching of evolution, not proponents of ID. As an example, he cites a recent panel discussion in which he participated featuring Chris Mooney, author of *The Republican War Against Science.* According to Arnhart, Mooney

> protested that high school students were not smart enough to read Darwin or to study the controversy over evolution. Instead, he insisted, they should only be presented with standard textbooks that

> summarize what the "experts" in biology believe. High school students should not be allowed to question those "experts."[215]

Arnhart responds that "[t]he mindless memorization of what the scientific 'experts' believe does not cultivate a serious intellectual ability to weigh scientific evidence and arguments."[216] He is right, but the problem with Mooney's position is even more basic. Mooney implies that the "experts" are unified on the evidence for neo-Darwinism. Yet there are a significant number of scientists who disagree with key parts of the theory, and they are beginning to make their voices known.

By 2006 more than 600 doctoral scientists had signed their names to "A Scientific Dissent from Darwinism,"[217] declaring they were "skeptical of claims for the ability of random mutation and natural selection to account for the complexity of life. Careful examination of the evidence for Darwinian theory should be encouraged." Signers of the declaration included members of the national academies of science in the United States, Russia, Poland, the Czech Republic, and India (Hindustan), as well as faculty and researchers from a wide range of universities and colleges, including Princeton, MIT, Dartmouth, the University of Idaho, Tulane, and the University of Michigan.

"Some defenders of Darwinism embrace standards of evidence for evolution that as scientists they would never accept in other circumstances," said signer Henry Schaeffer, director of the Center for Computational Quantum Chemistry at the University of Georgia.[218] Other signers expressed similar concerns. "The ideology and philosophy of neo-Darwinism, which is sold by its adepts as a scientific theoretical foundation of biology, seriously hampers the development of science and hides from students the field's real problems," declared Vladimir L. Voeikov, Professor of Bio-organic Chemistry at Lomonosov Moscow State University.[219] Microbiologist Scott Minnich at the University of Idaho complained that Darwinian theory was "the exceptional area that you can't criticize" in science, something he considered "a bad precedent."[220] Evolutionary biologist Stanley Salthe argued that biology students needed to be exposed to the weaknesses as well as the strengths of Darwin's theory:

> Darwinian evolutionary theory was my field of specialization in biology. Among other things, I wrote a textbook on the subject thirty years ago. Meanwhile, however I have become an apostate from Darwinian theory and have described it as part of modernism's origination myth. Consequently, I certainly agree that biology students at least should have the opportunity to learn about the flaws and limits of Darwin's theory while they are learning about the theory's strongest claims.[221]

Scientists skeptical of Darwin's theory are increasingly expressing their views during public debates on how evolution should be taught in schools. When the Ohio State Board of Education adopted a model lesson plan in 2004 on the "Critical Analysis of Evolution," two state university biologists, Daniel Ely and Glen Needham, publicly testified in favor of it.[222] Needham, a professor at the Ohio State University, argued that the model lesson "provides a corrective to the overly simplistic presentations that one often finds in high-school biology textbooks."[223] In Needham's view, the lesson plan was simply a matter of full disclosure to students of the current findings of biology. "The new lesson is good science because it encourages students to apply critical-thinking skills to analyze the evidence. Good students know there is a growing body of published criticism of evolutionary theory," he argued.

Darwin's scientific critics have also been vocal in other states. When the Kansas State Board of Education held hearings in 2005 on its new science standards on evolution, it heard testimony from what the *New York Times* called a "parade of Ph.D.s"[224] critical of Darwinian theory, including a professor of biochemistry and molecular biology from the University of Georgia, a geneticist from Cornell University, and a professor of microbiology from the University of Wisconsin-Superior.[225] When the South Carolina Board of Education considered a new science standard on the critical analysis of evolution in 2006, it heard testimony in support of the standard from evolutionary biologist Richard Sternberg, a research scientist at the National Institutes of Health and a Research Associate at the Smithsonian Institution's Museum of Natural History.[226] And after a federal judge struck down a textbook sticker in Cobb County, Georgia schools that encouraged students to approach their textbook's presentation of evolution "with an open mind," several dozen scientists, including faculty from the University of Georgia, Georgia Institute of Technology, and University of Alabama, filed an amicus brief with the court of appeals documenting "the current debate within scientific disciplines over whether chemical and biological (i.e., neo-Darwinian) evolution can adequately account for the origin of life and the development of life into its current forms."[227] (The brief filed by these scientists should be read by anyone who doubts there are mainstream scientific controversies surrounding Darwinian evolution. The amicus brief is reprinted at the end of this book as an appendix.)

The growing number of scientists who are coming "out of the closet" to express their skepticism of neo-Darwinism has made it harder for Darwinists to deny outright the existence of real scientific controversies involving evolution. Such denials become even more difficult when policymakers are presented with articles from the peer-reviewed science literature raising questions about such issues as the role of microevolution in explaining macroevolution, the efficacy of natural selection, the origin of animal body plans during the Cambrian explosion, or pointing out flaws in standard evidences of evolution such as vertebrate embryos and peppered moths.[228] These criticisms made of neo-Darwinism extend well beyond scientists who subscribe to intelligent design. As Discovery Institute points out, "many scientists critical of Darwinian evolution are also critical of intelligent design. If all the ID theorists in the world fell off the earth tonight, scientists tomorrow would still be pointing out weaknesses in Darwin's theory."[229]

Perhaps the most striking part of the evolution debate today is not that Darwinists promote the uncritical acceptance of experts, but that many Darwinists want to prevent students from learning about even scientific controversies that Darwinists themselves concede are legitimate.

This became evident during the 2003 debate over biology textbooks in Texas. Critics of Darwinian theory sought to correct factual errors about evolution in the textbooks as well as to add coverage regarding some of the scientific controversies over evolutionary theory. At the start of the debate, Texas Citizens for Science leader Steven Schafersman claimed that were no genuine scientific controversies over evolution for textbooks to cover.[230] But as public discussion progressed and critics of the textbooks enumerated specific examples of scientific controversies involving evolutionary theory, Schafersman's rhetoric began to, well, evolve.

Schafersman soon conceded that there are indeed some legitimate scientific controversies involving evolution, but asserted that these "[d]isagreements and controversies are found at the frontiers of research and graduate education, not at the level of introductory biology textbooks."[231] By the time the textbooks were actually voted on by the Texas State Board of Education, Schafersman was acknowledging that there are in fact "many disagreements among scientists about the correct nature or explanation of the evolutionary process."[232] According to Schafersman, these legitimate controversies included "[t]he sufficiency of microevolution to explain macroevolution," "[d]isagreements about the primacy of natural selection," and "[t]he extent to which evolutionary theory can explain or account for human morality, religion, behaviors, self-awareness, free will, etc." Notably, these were some of the very controversies that critics of the textbooks had been trying to raise.

Schafersman was reduced to arguing that high school students are too immature to appreciate them. "Scientific theories are too massive and established to expect any high school student to critique or question," he insisted. "The vast majority of high school students would not be able to perform such critiques in a scientific way. Scientific theories should be accepted as reliable knowledge in K–12 classes, and not made the object of questioning until they have the educational training necessary to do so, which consists of years of graduate study at universities." That is why even "[r]eal scientific problems, controversies, etc., should not be included in introductory science textbooks, because they are almost always too difficult to understand and their presence would only lead to student confusion and frustration."

Whatever the merits of this line of argument, it is entirely different than the standard sound bite that there are no legitimate scientific criticisms of evolution to cover. Sometimes when pressed about specific scientific controversies they cannot deny, the public partisans of evolution will insist that there is no debate over what they call "the fact of evolution," but then concede that there are still genuine debates over the "mechanism" of evolution. This is a rhetorical strategy at least as old as the Scopes trial. In order to deflect attention from the vigorous disagreements that exist even among evolutionists, public defenders of Darwin pretend as if debates over the mechanism of evolution are somehow inconsequential. But the mechanism of evolution is what *Darwinian* evolution is all about. Darwin purported to finally have an explanation for how an evolutionary process could actually occur. To concede that there is a scientific debate about the mechanism of evolution is really to concede that Darwin's theory itself is scientifically controversial, because Darwinian theory is primarily a claim about the mechanism of evolution. Presumably many evolutionists understand this fact, and that is why they so strongly oppose efforts to expose students to disagreements over Darwin's mechanism of natural selection.

Having said all of this, there remains the question of intelligent design. Even if there are legitimate scientific criticisms of Darwin's theory, that does not make intelligent design a convincing alternative theory to Darwin's undirected mechanism. Intelligent design must make its own case based on the scientific evidence. Arguing that it has not done so, Larry Arnhart raises two main objections: First, he criticizes intelligent design as purely negative argument from ignorance. Second, he faults intelligent design proponents for illegitimately extrapolating from human design to divine design. Other conservatives have added a third criticism: They claim that while intelligent design may be an interesting philosophical argument, it is not "science." All three criticisms are misplaced.

Is intelligent design a purely negative argument from ignorance?

Arnhart stresses "the purely negative character of intelligent design reasoning,"[233] and faults ID proponents of engaging in "the rhetorical technique of negative argumentation from ignorance."[234] Boiled down, the intelligent design argument in his view claims that if we cannot currently imagine how a structure such as "the bacterial flagellum could have arisen by purely natural causes or by chance, then we must assume it was designed by some sufficiently powerful intelligent agent."[235] Arnhart points out that this conclusion does not necessarily follow because such structures could have evolved due to "natural causes that we have not yet discovered."[236] Charles Krauthammer issues a similar indictment against intelligent design, arguing that "[i]t is a self-enclosed, tautological 'theory' whose only holding is that when there are gaps in some area of scientific knowledge—in this case, evolution—they are to be filled by God."[237]

Contrary to Arnhart and Krauthammer, however, the primary argument for intelligent design is positive, not negative. Based on our own extensive experience in the natural and social worlds, we know that intelligent causes are habitually capable of producing certain kinds of highly functional complexity (what mathematician William Dembski defines as "specified complexity"[238]). Thus, whenever we find highly-functional complexity we have good *prima facie* evidence that it may have been produced by an intelligent cause—because we *know* from our own experience that intelligent causes habitually produce this kind of complexity. This argument is based on our knowledge of the cause and effect operation of the world around us, not on our ignorance. So when we see Mt. Rushmore, or a laptop computer, or even a beaver's dam and a bird's nest, we are acting rationally when we conclude that they were the result of purposeful actions rather than processes of chance and necessity.

Arnhart insists, however, that "for intelligent design to have some positive content," its proponents "would have to explain exactly where, when, and how a disembodied intelligence designed flagella and attached them to bacteria."[239] But why? Can't we know that something was produced by an intelligent cause even if we do not know the method used by the intelligent cause? If I am hiking in the desert and come across what appears to be the ruin of a giant building, can't I conclude that an intelligent cause was at work even if I know nothing about *how* the building was created or *who* had inhabited the area previously? Discovering the methods by which intelligent causes organized matter and energy in a certain way is an interesting inquiry, but determining whether an intelligent cause was involved at all also seems to be a legitimate question. Faulting intelligent design theory for not answering a question it does not even purport to answer seems unreasonable.

It might be added that the positive case for intelligent design is based on the same scientific methodology employed by Darwin himself. Darwin embraced what is known as "uniformitarianism," an approach popularized by his friend Charles Lyell, the founder of modern geology. According to uniformitarianism, science seeks to explain past events by invoking causes that regularly operate in nature today rather than by hypothesizing unique or *ad hoc* causes that may have existed in the past. As research chemist and intelligent design proponent Charles Thaxton explained in the late 1980s, uniformitarianism assumes that

> the causes we observe today are... uniform over space and time. We assume that whatever cause and effect relationships we experience today held in the past as well. If we observe certain effects today, and determine their causes by experience, then whenever we see evidence that the same kind of effects took place in the past, we are

> justified in assuming the same causes were at work. In other words we rely on uniform sensory experience to extend the assignment of causes into areas where we don't have experience.[240]

This was precisely the scientific approach followed by Darwin when he drew on his knowledge of the effects of breeding or "artificial selection" to provide evidence for his view that "natural selection" was the primary mechanism of evolution. The same uniformitarian approach is employed by the modern proponents of intelligent design who infer intelligent causation in the past based on what we know intelligent causes are capable of doing in the present.

Of course, conclusions reached by applying uniformitarianism can be mistaken. It is logically possible for similar effects to be the products of different causes. That is why the positive argument for intelligent design is strengthened by a negative argument: At the same time intelligent causes habitually produce certain kinds of highly-functional complexity, non-intelligent causes (chance and necessity) do not seem capable of generating these same kinds of highly-ordered complexity. This is the point of Michael Behe's argument about "irreducible complexity." According to Behe, a biological system is "irreducibly complex" if it "is composed of several interacting parts that contribute to the basic function" and if "the removal of any one of the parts causes the system to effectively cease function."[241] Irreducibly complex systems cannot be generated through a direct evolutionary route by natural selection "since any precursor to an irreducibility complex system is by definition nonfunctional. Since natural selection requires a function to select, an irreducibly complex biological system, if there is such a thing, would have to arise as an integrated unit for natural selection to have anything to act on."

Taken together, the positive and negative arguments offered by intelligent design scientists reinforce each other: If we know that intelligent causes habitually generate systems of highly-functional complexity, and we also know that non-intelligent causes typically do not do so, then the best explanation under the circumstances ("the inference to the best explanation"[242]) is that an intelligent cause was involved whenever we see highly-functional complexity. Intelligent design as a scientific hypothesis is thus a probabilistic argument based on our current knowledge, which means that it is also subject to testing and possible refutation based on new evidence of how nature operates.[243]

Although intelligent design is first and foremost a positive argument, Arnhart is right to see that the negative argument is quite important. If non-intelligent processes are typically capable of producing new highly-functional complexity, then it brings into question whether intelligent causes are the best explanation for any given case of such complexity. At the same time, the less capable non-intelligent causes are of producing highly-functional complexity, the more likely intelligent causation is the best explanation, given what we know about the capabilities of intelligent causes. But Arnhart finds the negative argument offered by intelligent design proponents wanting as well, focusing in particular on Michael Behe's concept of "irreducible complexity."

Arnhart is ultimately unconvinced by Behe's account of irreducible complexity, as is James Q. Wilson, who asserts that "teams of biologists" have shown it to be "false."[244] Wilson does not cite the unnamed biologists who are supposed to have made such a showing or describe their evidence, but Arnhart at least tries to offer a justification for his view. Drawing on the work biologist Kenneth Miller, Arnhart argues that "complex biochemical mechanisms can be assembled by natural selection working indirectly by combining simpler mechanisms that originally served some other function."[245]

This is known as the "co-option" argument—natural selection is envisioned as "co-opting" parts of existing biological systems in order to build completely new systems with completely different purposes. Behe, however, anticipated this objection in his book *Darwin's Black Box*. Agreeing that "one cannot definitely rule out the possibility of an indirect, circuitous route," he pointed out that "[a]s the complexity of an interacting system increases... the likelihood of such an indirect route drops precipitously."[246]

Since the mere *possibility* of indirect evolutionary pathways does not make such pathways *plausible*, Arnhart seeks to show that there is in fact evidence that indirect Darwinian evolution could produce irreducibly complex structures. Relying again on Kenneth Miller, Arnhart cites the Type Three Secretory System (TTSS), which he argues could have been a precursor of the bacterial flagellum, which has become known as Behe's star example of irreducible complexity:

> Some bacteria have a type III secretory system (TTSS) that allows them to inject protein toxins into the cells of host organisms. Some of the protein structures in the TTSS are remarkably similar to the protein structures in the flagellum, which suggests that the flagellum could have evolved by incorporating the structures of the TTSS, so that mechanisms originally serving one function could be taken up into new mechanisms serving new functions. It seems that if we remove some of the parts of the flagellar system, it will no longer perform the function of propelling the bacterium through water, but it might perform some other function such as that of the TTSS. We can then imagine a "circuitous route" to the evolution of the bacterial flagellum.[247]

But the claim that TTSS supplies a Darwinian precursor to the bacterial flagellum has been strongly challenged by University of Idaho microbiologist Scott Minnich and philosopher of science Stephen Meyer. As Minnich and Meyer point out, "phylogenetic analyses of the gene sequences suggest that flagellar motor proteins arose first and those of the pump came later. In other words, if anything, the pump evolved from the motor, not the motor from the pump."[248] Furthermore,

> the other thirty proteins in the flagellar motor (that are not present in the TTSS) are unique to the [flagellar] motor and are not found in any other living system. From whence, then, were these protein parts co-opted? Also, even if all the protein parts were somehow

> available to make a flagellar motor during the evolution of life, the parts would need to be assembled in the correct temporal sequence similar to the way an automobile is assembled in factory. Yet, to choreograph the assembly of the parts of the flagellar motor, present-day bacteria need an elaborate system of genetic instructions as well as many other protein machines to time the expression of those assembly instructions. Arguably, this system is itself irreducibly complex. In any case, the co-option argument tacitly presupposes the need for the very thing it seeks to explain—a functionally interdependent system of proteins.[249]

In sum, the Type Three Secretory System offers no credible evidence that the selection-mutation mechanism can produce irreducibly complex structures.

It is worth noting that although public defenders of Darwinism such as Kenneth Miller have attempted to debunk Behe's criticisms of the evidence for natural selection, some prominent biochemists have conceded them. Shortly after Behe's *Darwin's Black Box* came out in 1996, biochemist James Shapiro of the University of Chicago acknowledged that "there are no detailed Darwinian accounts for the evolution of any fundamental biochemical or cellular system, only a variety of wishful speculations."[250] Five years later in a scientific monograph published by Oxford University Press, biochemist Franklin Harold, who rejects intelligent design, admitted in virtually the same language: "We must concede that there are presently no detailed Darwinian accounts of the evolution of any biochemical or cellular system, only a variety of wishful speculations."[251]

Other scientists have begun to cite Behe's ideas favorably in their own scientific publications. In a 2001 technical article in *Nature's* peer-reviewed *Encyclopedia of the Life Sciences*, two biologists cited Behe's identification of "irreducibly complex structures," as evidence for the possible limits of the mutation-selection mechanism, noting that "[u]p to now, none of these systems [described by Behe] has been satisfactorily explained by neo-Darwinism."[252] Similarly, a 2002 article published in the peer-reviewed *Annals of the New York Academy of Sciences* stated that while "Michael Behe may have been overly hasty in dismissing the possibility of the evolution of such mechanisms by natural selection... his notion of 'irreducible complexity' *surely captured a feature of developmental systems that is of major importance.*"[253]

Moreover, irreducible complexity is only one of the objections to the sufficiency of the selection-mutation mechanism to account for the biological complexity we see in nature. As Stephen Meyer notes, "[m]any scientists and mathematicians have questioned the ability of mutation and selection to generate information in the form of novel genes and proteins. Such skepticism often derives from consideration of the extreme improbability (and specificity) of functional genes and proteins."[254] Accordingly, Michael Behe and David Snoke published evidence showing the limits of unguided evolutionary processes to produce functional novel proteins in the peer-reviewed journal *Protein Science*,[255] and Douglas Axe has reported experimental evidence in the peer-reviewed *Journal of Molecular Biology* showing the extreme rarity of functional proteins among possible amino acid sequences, which casts doubt on the ability of random variation to generate such proteins.[256] Nor are criticisms of the selection-mutation mechanism limited to the proponents of intelligent design. There are many scientists who reject intelligent design who offer similar criticisms of the Darwinian mechanism. University of Massachusetts geneticist Lynn Margulis, for example, declares bluntly that "new mutations don't create new species; they create offspring that are impaired."[257] Suffice to say, negative arguments against the Darwinian mechanism have not been refuted.

Does ID illegitimately extrapolate from human to divine design?

According to Arnhart, "a fundamental flaw" of intelligent design theorists is "their equivocal use of the term 'intelligent design.'" In his view, ID proponents offer evidence based on "*human* 'intelligent design,'" and then try to extrapolate it to prove "*divine* 'intelligent design.'"[258] Arnhart is not the only conservative to fault ID proponents for trying to extrapolate to the supernatural. While praising ID theorists for highlighting Darwinism's difficulties, noted bioethicist Leon Kass has argued that "the IDers' assertion that the only possible answer is a Designer-God is not warranted. There is simply no evidence in support of this proposition."[259] George Will, Charles Krauthammer, and James Q. Wilson similarly fault intelligent design for trying to prove the existence of God through empirical science, while John Derbyshire of *National Review* cites with approbation a reader's view that intelligent design is illegitimate because it attributes the world to "supernatural forces" and "[s]upernatural forces are simply not within the scope of science."[260] These conservatives all seem to agree with Judge John Jones in the *Kitzmiller v. Dover* case that intelligent design "requires supernatural creation."[261]

However, such criticisms are based on almost complete ignorance of what intelligent design theorists actually propose. Arnhart argues that "insofar as we have never directly observed a disembodied, omniscient, and omnipotent intelligence causing effects that are divinely designed, we cannot infer a divine intelligent designer fom our common human experience."[262] Arnhart does not seem to realize that ID proponents actually agree with him. They have consistently argued that empirical science—by itself—cannot determine whether an intelligent cause detected through science is inside or outside of "nature," let alone whether the intelligent cause is the Judeo-Christian God who is "omniscient... and omnipotent." As early ID proponent Charles Thaxton cautioned in the 1980s, "[b]y experience we cannot determine whether the inferred intelligent cause is within the universe (Naturalism) or beyond it (Supernaturalism)."[263]

This point has been a repeated refrain in both the technical and popular writings of ID proponents for the past 17 years.[264] For example, in the first two editions of the early intelligent design textbook *Of*

Pandas and People (originally issued in 1989), one finds the following statements:

> If science is based upon experience, then science tells us the message encoded in DNA must have originated from an intelligent cause. What kind of intelligent agent was it? *On its own, science cannot answer this question; it must leave it to religion and philosophy.* But that should not prevent science from acknowledging evidences for an intelligent cause origin wherever they may exist.[265]

> [T]he place of intelligent design in science has been troubling for more than a century. That is because on the whole, scientists from within Western culture failed to distinguish between intelligence, which can be recognized by uniform sensory experience, and the supernatural, which cannot. Today we recognize that appeals to intelligent design may be considered in science, as illustrated by the current NASA search for extraterrestrial intelligence (SETI). Archaeology has pioneered the development of methods for distinguishing the effects of natural and intelligent causes. *We should recognize, however, that if we go further, and conclude that the intelligence responsible for biological origins is outside the universe (supernatural) or within it, we do so without the help of science.*[266]

One can find similar statements in the writings of Michael Behe and William Dembski. Distinguishing the modern scientific theory of intelligent design from the earlier "natural theology" of William Paley, for example, Behe has stressed:

> The most important difference [between modern intelligent design theory and Paley's arguments] is that [intelligent design] is limited to design itself; I strongly emphasize that it is not an argument for the existence of a benevolent God, as Paley's was. I hasten to add that I myself do believe in a benevolent God, and I recognize that philosophy and theology may be able to extend the argument. But a scientific argument for design in biology does not reach that far.[267]

Ironically, ID proponents are often criticized for insisting that a design inference within science cannot determine whether an intelligent cause is "supernatural." Because many ID theorists believe in God—just like the vast majority of Americans—they are accused of being "disingenuous" or "dishonest" when they insist on the limited nature of the design inference in science. According to critics, proponents of ID falsely claim that they don't know "who the intelligent designer is" when in reality they believe the designer to be God. But this is a caricature of what ID proponents actually say. I don't know of any leading ID proponent who hides whether or not he believes in God. Charles Thaxton, Michael Behe and William Dembski are certainly open about their religious convictions. The relevant question is not whether proponents of ID personally believe in God, *but whether they believe empirical science can identify an intelligent cause operating in nature as God*. Clearly, they do not believe this. Instead of being disingenuous, ID proponents' insistence that the design inference in science cannot determine whether an intelligent cause is "supernatural" is simply an effort to be honest about the limits of modern science.

Yet Arnhart appears to agree that ID proponents are being disingenuous on the question of God. After quoting William Dembski insisting that intelligent design "involves no recourse to the supernatural," he supplies additional quotes from Dembski claiming that "intelligent design is just the *Logos* theology of John's Gospel restated in the idiom of information theory" and that "Christ is indispensable to any scientific theory." Arnhart comments that "[h]ere the 'recourse to the supernatural' is clear," leaving the impression that Dembski is being hypocritical.[268] But the quotes cited by Arnhart are taken out of context and misrepresent Dembski's real position. One of the quotes comes from Dembski's 1999 book *Intelligent Design: The Bridge Between Science and Theology*. Inexplicably, Arnhart neglects to cite Dembski's explicit discussion in that book of the limits of the design inference:

> Intelligent design is modest in what it attributes to the designing intelligence responsible for the specified complexity in nature. For instance, design theorists recognize that the nature, moral character and purposes of this intelligence lie beyond the remit of science... [Thus,] [i]ntelligent design as a scientific theory is distinct from a theological doctrine of creation. Creation presupposes a Creator who originates the world and all its materials. Intelligent design attempts only to explain the arrangement of materials within an already given world. Design theorists argue that certain arrangements of matter, especially in biological systems, clearly signal a designing intelligence... Intelligent design requires an intelligent cause that is capable of arranging complex specified structures. That capacity to arrange matter, however, is exercised within space and time, and need not violate any laws of nature. Intelligent design does not require a creator that originates the space, time, matter and energy that together constitute the universe.[269]

To say that intelligent design as a scientific theory cannot determine whether an intelligent cause behind life is "supernatural" does not mean that ID is irrelevant to debates over the existence of God. Just as Darwinism has anti-theistic implications, the theory of intelligent design may have implications favorable to theism. A belief that life is the product of an intelligent cause is certainly more harmonious with traditional theism than a belief that life is the product of an undirected process of chance and necessity. Consequently, philosophers and theologians friendly to traditional theism may well draw on the insights of intelligent design to develop their arguments in support of theism—just as philosophers and theologians hostile to traditional theism have drawn on the insights of Darwinian theory to debunk traditional religious beliefs. But these implications of intelligent design for other disciplines are distinct from the question of whether the design inference in science logically necessitates the existence of a supernatural creator. Clearly, it does not.

Is intelligent design philosophy rather than science?

Some conservatives are willing to concede that intelligent design may be true, but they contend that it is "philosophy" rather than "science." The proper definition of science has been a contentious issue among scholars, and this small volume obviously cannot settle the controversy.[270] However, it is worth pointing out that the objection cuts both ways. Evolution by natural selection was offered by Darwin as a scientific disproof of the idea that the development of life was guided by an intelligent cause. In other words, Darwin and his followers thought they had produced scientific evidence refuting design in nature. But if Charles Darwin, Richard Dawkins, and other biologists can offer scientific evidence they think refutes design, then why does it suddenly become "unscientific" for Michael Behe to highlight scientific evidence he believes supports design? Evolutionists cannot have it both ways. Either Darwinism and intelligent design are both philosophy, or they are both science. Both theories explicitly try to explain the "appearance of design" observed throughout nature. Both study present-day processes and then attempt to apply those processes to the historical record to determine how best to account for the observed data. As philosopher of science Stephen Meyer points out, the two theories are methodologically equivalent.[271] Unfortunately, most critics of intelligent design rarely apply their definition of science consistently.

Because Darwinism and design are both "historical sciences," neither is open to the same level of empirical investigation and proof that is found in experimental sciences such as chemistry. But that is simply the nature of the case. As explained by eminent evolutionary biologist Ernst Mayr,

> the evolutionist attempts to explain events and processes that have already taken place. Laws and experiments are inappropriate techniques for the explication of such events and processes. Instead one constructs a historical narrative, consisting of a tentative reconstruc-

tion of the particular scenario that led to the events one is trying to explain.[272]

Regardless of whether one believes that "historical sciences" are properly "scientific," attempts to artificially constrain the debate over evolution by allowing only one side of the debate to be heard are unhelpful. Indeed, they undercut science as a genuine search for the truth about the natural world.

CONCLUSION

"Darwinism... is the 'politically correct' of science."
Italian geneticist Giuseppe Sermonti.[273]

The allure of Darwinian conservatism is not hard to understand. While nineteenth century giants such as Karl Marx and Sigmund Freud have been debunked, Darwin retains his prestige among the elites as a secular saint. Moreover, Darwinists have clothed themselves in the mantle of modern science, successfully stigmatizing those who criticize them as bigoted Bible-thumpers who are "anti-science."

No wonder a number of conservative intellectuals either refrain from becoming involved in the debate over Darwinism or take the side of Darwin as a matter of course. It is regarded as unfashionable or even embarrassing to be on the side of Darwin's critics, and who wants to be unfashionable or embarrassed?

One suspects that this concern for being fashionable has something to do with the dismissive attitude toward intelligent design taken by conservative pundits such as George Will and Charles Krauthammer, neither of whom shows evidence of having read or considered the arguments made by intelligent design proponents. If they had, they would not assert so tritely that intelligent design is "warmed-over creationism"[274] or an attempt "to compel public education to infuse theism into scientific education."[275] Nor would Krauthammer have denounced the Kansas State Board of Education for "forcing intelligent design into the statewide biology curriculum"[276] when the Board made clear it had done the exact opposite: "We also emphasize that the Science Curriculum Standards do *not* include Intelligent Design."[277] Which part of the phrase "do not include Intelligent Design" did Krauthammer fail to understand? Sadly, I doubt he even bothered to read the Kansas science standards.[278] It is ironic that some conservatives who would not trust left-wing reporting, say, about the war in Iraq, will apparently accept wholesale anything the media report about the controversy over evolution. Because pundits like Will and Krauthammer have not done their homework, they are reduced to repeating such canards as the claim that intelligent design is "not falsifiable"[279]— not realizing that intelligent design makes falsifiable predictions in the same way that Darwinian theory does, as has been explained with some care by a number of intelligent design theorists.[280]

Other, more thoughtful, conservatives remain troubled by what they regard as the excesses of Darwinian ideologues, but they seem to think they can neutralize Darwinism by redefining it. For example, physicist Stephen Barr has argued in *First Things* that neo-Darwinism, properly understood, need not require a process that is "unguided" or "unplanned." "The word 'random' as used in science does not mean uncaused, unplanned, or inexplicable; it means uncorrelated," he insists.[281] The problem is not that Barr is wrong about the appropriate meaning of "random," but that mainstream Darwinists do not accept his point. As pointed out in the introduction, Darwinism from the start has been defined as an undirected process. That is its core, and that is why Darwin himself emphasized that "no shadow of reason can be assigned for the belief that variations... were intentionally and specially guided."[282] Barr may be correct that a more modest Darwinism that does not insist on evolution being undirected would be harmless, but the more salient point is that it would no longer be Darwinism. Conservatives cannot resolve the problems with Darwinian evolution merely by offering their own idiosyncratic definition of the term.

Still other conservatives, like Larry Arnhart and James Q. Wilson, believe that properly understood, Darwin's theory can be used to support moral universals and temper utopian schemes. But their argument flies in the face of an historical record that demonstrates the opposite.

For the past hundred years, mainstream Darwinists have drawn on Darwin's theory to promote relativism and utopian social reforms such as eugenics. Of course, these mainstream Darwinists could have been wrong, but it seems to me that a strong case can be made that their efforts were logically connected to Darwin's theory. If one believes that all human behaviors are equally the products of natural selection, and that they all exist only because they somehow promote biological survival, it is hard to see an objective ground for condemning any particular behavior. The maternal instinct is natural according to Darwin, but so is infanticide. Monogamy is natural, but so is polygamy. If a certain man prefers ten wives to one, who are we to judge? Obviously natural selection has preserved the desire for multiple wives in that male, so polygamy must be "right" for him. Carson Holloway is right to conclude that "[i]n the absence of some principles of goodness that transcend our nature as Darwinism presents it, there can be no intelligible reason to prefer our noblest but weakest desires to our strongest and most commonplace ones, [or] to prefer our strong and decent desires to our equally strong but unscrupulous ones."[283]

Similarly, if one believes that human progress is dependent on a vigorous struggle for existence, then any diminishment of natural selection in human society will raise legitimate concerns, and efforts to reinstate selection through eugenics may well appear rational. In addition, once one understands the evolving nature of "human nature," it is difficult to see any *in principle* objection to efforts to transform human nature through bioengineering. Drawing on a report from the President's Council on Bioethics, Arnhart attempts to outline a Darwinian argument against radical human bioengineering, but his argument is less than persuasive. "Our desires have been formed by natural selection over evolutionary history to promote survival and reproduction," he writes.[284] "Knowing this should make us cautious about using biotechnology to radically change our evolved nature." But why? Natural selection is a messy, hit-or-miss process of dead-ends and false starts. Why shouldn't human beings use their reason to direct their evolution in order to produce a new kind of human being? What is so sacrosanct about existing human dispositions and capacities, since they were produced by such an imperfect and purposeless process?

Arnhart and the President's Council on Bioethics seem to want to clothe human nature with a kind of sacred awe that will restrain human beings from tinkering with it. But such awe is alien to the Darwinian mindset. In his autobiography, Darwin recounted how he had once had such feelings, but they had evaporated:

> In my Journal I wrote that whilst standing in the midst of the grandeur of a Brazilian forest, "it is not possible to give an adequate idea of the higher feelings of wonder, admiration, and devotion which fill and elevate the mind." I well remember my conviction that there is more in man than the mere breath of his body; but now the grandest scenes would not cause any such convictions and feelings to rise in my mind.[285]

In the Darwinian framework, there is nothing intrinsically right about the current capacities of human beings, so there can be nothing intrinsically wrong about trying to alter them. In the end, Arnhart's main arguments against radical human bioengineering are his prediction that it may not be technically feasible and his hope that it may be restrained by certain deeply-ingrained human desires. Let's certainly hope so, but Darwinism itself provides little or no barrier against such schemes. As Carson Holloway points out, the Darwinian account of morality all but invites "wholesale biological engineering."[286]

Despite my fundamental disagreement with Darwinian conservatives such as Arnhart and Wilson, I applaud them for getting one key issue right: They understand that Darwinism has far-reaching implications for society, for morality, and for politics. Unlike conservatives who try to ignore the debate over evolution, these Darwinian conservatives recognize that modern evolutionary theory raises important questions that need to be addressed by all serious citizens.

Conservatives who would rather sit out the evolution controversy need to understand that the current debate is not primarily about religious fundamentalism, nor is it simply a rehashing of certain esoteric points of biology and philosophy. Contrary to *National Review's* Jonah Goldberg, the debate over Darwin is not merely "an abstract battle which saps energy from more important issues."[287] Darwinian reductionism has become culturally pervasive, becoming inextricably intertwined with contemporary conflicts over traditional morality, personal responsibility, sex and family, and bioethics. Indeed, Darwinian reductionism ultimately undermines the idea of man as a rational being who is both free and accountable. As Harry Jaffa has argued, "[t]he attempt, inherent in Darwinism, to reduce man's humanity to the outcome of a blind struggle of material forces denies, in effect, the ground in nature and reason of man's personal and political freedom."[288]

Darwinism is also central to an important debate about the role of scientific expertise in American society that dates backs to the Progressive era. Darwin's defenders have been at the forefront of promoting technocracy—the claim that scientific experts ultimately have the right to rule free from the normal restraints of democratic accountability. Disparaging the wisdom of ordinary citizens and their elected representatives, dogmatic Darwinists essentially argue that public policy should be dictated by the majority of scientific experts without input from anyone else. Today this bold assertion is made not just with regard to evolution, but concerning a host of other controversial issues such as sex education, euthanasia, embryonic stem-cell research, cloning, and global warming. Those on the left declare that any dissent from liberal orthodoxy on these issues represents a "war on science."[289]

The effort to demonize normal democratic dissent in the area of science and public policy has been fomented by Fenton Communications, the far-left public relations firm for such groups as MoveOn.org, Planned Parenthood, the American Trial Lawyers Association, Greenpeace, and the National Abortion Rights Action League (NARAL).[290]

With funding from the Tides Center, Fenton has set up a group bearing the Orwellian name of the "Campaign to Defend the Constitution" ("DefCon").[291] According to DefCon, good science just happens to equal the political agenda of the left, and anyone who says otherwise is a theocrat who opposes "scientific progress."

The effort to deny ordinary citizens their voice in public policy in the name of "scientific progress" has deep roots in the Social Darwinism of the past. Speaking before the SecondInternational Congress of Eugenics in 1921, Alleyne Ireland declared that modern conditions had rendered America's original form of government established by the Constitution and the Declaration of Independence "utterly unsuitable."[292] America's founders believed that "governments derive their just powers from the consent of the governed," and they set up arrangements "designed with a view to making abuse of power difficult." But in an age when government must increasingly provide a wide range of social services, society could no longer afford to rely on government by non-experts according to Ireland. Instead, it was "imperative... that the omnipresent activity of government should be guided by the light of scientific knowledge and conducted through the instrumentality of a scientific method." Most contemporary boosters of technocracy are not quite as blunt as Ireland, but their overall philosophy is the same.

Some conservatives seem all too tempted by this idea that scientists rather than ordinary people should rule. *National Review's* John Derbyshire, for example, has claimed that "[i]f only Ph.D.s in the sciences had the vote, we should have Libertarian govt.," clearly believing that society would be a better off if scientists were allowed to govern free from interference by ordinary citizens.[293] While Derbyshire later retracted this claim after being challenged, his apparently boundless trust in scientific expertise is breathtaking.

Of course, there is much that can be said in favor of the authority of scientific expertise in modern life. In an increasingly complex and technologically-driven world, the need for scientific input on public policy would seem obvious. Since many policy questions today arise in such science-based fields as medicine, transportation, and ecology, why shouldn't politicians and voters simply defer to the authority of scientific experts in these areas?

While this line of reasoning exhibits a surface persuasiveness, it ignores the natural limits of scientific expertise. Scientific knowledge may be necessary for good public policy in certain areas. But it is not sufficient. Political problems raise questions of justice, equity, and prudence, and scientists are ill-equipped to function as the judges of such questions. British writer C. S. Lewis warned about this drawback of technocracy back in the 1950s. "I dread specialists in power, because they are specialists speaking outside their special subjects," Lewis wrote. "Let scientists tell us about sciences. But government involves questions about the good for man, and justice, and what things are worth having at what price; and on these a scientific training gives a man's opinion no added value."[294]

Technocracy poses a further difficulty: Experts can be wrong, sometimes egregiously. If the history of "Social Darwinism" in politics shows anything, it is that scientific experts can be as fallible as anyone else. They are capable of being blinded by their own prejudices and going beyond the evidence in order to promote the policies they favor. Alfred Kinsey's empirical claims about the sexual behavior of the general American public were junk science, given his deeply flawed sample population; yet that did not stop him from boldly making his claims and vigorously defending them as sound science.

What is true of individual scientists can be true of the scientific community as a whole. For decades eugenics was embraced as legitimate by America's leading scientists and scientific organizations such as the American Association for the Advancement of Science. Critics of eugenics, meanwhile, were roundly stigmatized as anti-science and religious zealots. Yet the critics were the ones who turned out to be right, not the scientific elites. Darwinian conservatives today forget the lessons of history and sometimes display astonishing naïveté about the trustworthiness of current scientific elites.

James Q. Wilson, for example, appeals to the sacral authority of the American Association for the Advancement of Science to justify his judgment that intelligent design is not science.[295] The AAAS board issued a resolution condemning intelligent design in 2002. What Prof. Wilson apparently does not know is that the AAAS board members who approved this resolution did not even bother to study the arguments for themselves. After the resolution was issued, I surveyed AAAS board members about what articles and books they had read by proponents of intelligent design. Of the four board members who responded, none could identify even a single article or book. While one board member said vaguely that she had perused unspecified sources on the internet, Alan Leshner, the head of the AAAS, couldn't even make that claim. He responded that the issue had been looked at by the group's "science policy staff." Now there is a good way to determine the validity of a new scientific theory: Don't investigate the evidence for yourself. Does Prof. Wilson wish to hold this up as a model for how the "scientific community" should determine scientific truth? Does he think it is good science (or even science at all) to condemn a new scientific idea without even bothering to read those scientists who are proposing it?

The bottom line is that it is extremely valuable in a democratic society for public policy claims made by scientists to be scrutinized by policymakers and citizens just as much as public policy claims made by other interested parties. Any suggestion that policymakers should simply rubber-stamp the advice of the current majority of scientists is profoundly subversive of the principles of representative democracy and American republicanism. Free governments are supposed to represent the interests of all citizens, not just members of a scientific elite who may be blinded by their own ideology. As equal citizens before the law, scientists have every right to inform policymakers of the scientific implications of their actions. But they have no special right to demand that policymakers listen to them alone.

Even conservatives who accept Darwinian theory should think twice before embracing the dogmatic claims to rule made by Darwinists and other scientific materialists. Such claims have resulted in a concerted effort to shut down honest debate through caricatures and intimidation. While evolutionists continue to portray themselves as the victims of fundamentalist intolerance, in most places today it is the evolutionists who have turned inquisitors, and it is the critics of Darwin's theory who are being persecuted.

At George Mason University in Virginia, biology professor Caroline Crocker made the mistake of favorably discussing intelligent design in her cell biology class. She was suspended from teaching the class, and then her contract was not renewed.[296]

At the Smithsonian Institution, evolutionary biologist Richard Sternberg, the editor of a respected biology journal, faced retaliation by Smithsonian executives in 2005 after accepting for publication a peer-reviewed article favoring intelligent design. Investigators for the U.S. Office of Special Counsel later concluded that "it is... clear that a hostile work environment was created with the ultimate goal of forcing [Dr. Sternberg]... out of the [Smithsonian]."[297]

When asked about Sternberg's plight by *The Washington Post*, Eugenie Scott of the National Center for Science Education seemed to suggest that Sternberg was lucky more wasn't done to get rid of him: "If this was a corporation, and an employee did something that really embarrassed the administration, really blew it, how long do you think that person would be employed?"[298]

The same burn-them-at-the-stake approach is being applied to scientists who criticize Darwin without raising the issue of intelligent design. At the Mississippi University for Women, chemistry professor Nancy Bryson was removed as head of the division of natural sciences in 2003 after merely presenting scientific criticisms of biological and chemical evolution to a seminar of honors students. "Students at my college got the message very clearly, do not ask any questions about Darwinism," she explained later.[299]

Students at other educational institutions are receiving the same chilling message. In 2005, Ohio State University doctoral candidate Bryan Leonard had his dissertation defense placed on hold after three pro-Darwin professors filed a bogus complaint attacking Leonard's dissertation research as "unethical human subject experimentation." Leonard's dissertation project looked at how student beliefs changed after students were taught scientific evidence for and against modern evolutionary theory. The complaining professors admitted that they had not actually read Leonard's dissertation. But they were sure it must be unethical. Why? According to the professors, there is no valid evidence against evolutionary theory. Thus—by definition—Leonard's research must be tantamount to child abuse.[300]

These politically-correct efforts to purge the educational system of any critics of Darwin are fueled by increasingly toxic rhetoric on the part of evolutionists. Rather than defend the scientific merits of evolution, Darwinists have become obsessed with denouncing their opponents as dangerous zealots hell-bent on imposing theocracy.

In many states, it has become routine to apply the label of "Taliban" to anyone who supports teaching students about scientific criticisms of Darwinian theory.[301] Biology professor P. Z. Myers at the University of Minnesota, Morris, has even demanded "the public firing and humiliation of some teachers" who express their doubts about Darwin.[302] He further says that evolutionists should "screw the polite words and careful rhetoric. It's time for scientists to break out the steel-toed boots and brass knuckles, and get out there and *hammer* on the lunatics and idiots."[303] Defenders of evolution who claim to fear blind zealotry might want to look in the mirror. The new Darwinian fundamentalists have become just as intolerant as the religious fundamentalists they despise.

Such intolerance should raise concerns for people from across the political spectrum. True liberals—those who favor free and open debate—should be appalled by the growing campaign of intimidation against academic critics of Darwinism just as much as conservatives. Whatever one's personal view of Darwinism, the current atmosphere is unhealthy for science, and it's unhealthy for a free society.

Conservatives who are discomfited by the continuing debate over Darwin's theory need to understand that it is not about to go away. It is not going away because the accumulating discoveries of modern science undercut rather than confirm the claims of neo-Darwinism. It is not going away because Darwinism fundamentally challenges the traditional Western understanding of human nature and the universe. Finally, it is not going away because free men and women do not like to be told that there are some questions they are not allowed to ask, and there are some answers they are not allowed to question.

The debate over Darwin is not a sideshow. It is central to arguments over moral relativism, personal responsibility, limited government, and scientific utopianism. If conservatives want to address root causes rather than just symptoms, they need to join the debate, not scorn it or ignore it.

APPENDIX A

Brief of Amici Curiae Biologists And Other Scientists In support of Appellants in the case of Cobb County School District v. Selman

In 2002 Georgia's Cobb County School District required stickers to be affixed to the district's biology textbooks stating: "This textbook contains material on evolution. Evolution is a theory, not a fact, regarding the origin of living things. This material should be approached with an open mind, studied carefully, and critically considered." Alleging that the stickers violated the First Amendment's ban on establishments of religion, the American Civil Liberties Union (ACLU) sued the school district. In January 2005 Judge Clarence Cooper struck down the stickers as unconstitutional, but in May 2006 the U.S. Court of Appeals for the Eleventh Circuit vacated Cooper's ruling and sent the case back to the district court in order to clear up a confused factual record.

The following amicus brief was submitted to the federal appeals court by nearly fifty scientists who were concerned that the ACLU was attempting to censor legitimate discussion of scientific criticisms of Darwinian evolution. The brief, whose signatories include science professors at the University of Georgia, Georgia Institute of Technology, Emory

University, and the University of Florida, presents a succinct description of the many scientific controversies surrounding biological and chemical evolution documented in the standard scientific literature.

IN THE UNITED STATES COURT OF APPEALS
FOR THE ELEVENTH CIRCUIT

No. 05-10341-II

Cobb County School District, et al.,
Defendants/Appellants

v.

Jeffrey Michael Selman, et al.,
Plaintiffs/Appellees,

BRIEF OF AMICI CURIAE BIOLOGISTS AND OTHER SCIENTISTS IN SUPPORT OF APPELLANTS

GEORGE M. WEAVER*
KEVIN T. McMURRY
Hollberg & Weaver, LLP
2941 Piedmont Road, N.E. – Suite C
Atlanta, GA 30305

SETH L. COOPER
Center for Science & Culture
Discovery Institute
1511 Third Ave., Suite 808
Seattle, WA 98101

* Counsel of Record

Selman v. Cobb County School District, No. 05-10341-II

CERTIFICATE OF INTERESTED PERSONS AND CORPORATE DISCLOSURE STATEMENT

Undersigned counsel of record for Amici Curiae Biologists and other Scientists hereby certifies that the following persons and entities have an interest in the outcome of this case:

Appellants/Defendants:

Cobb County Board of Education

Cobb County School District

Joseph Redden, Superintendent

Counsel for Appellants/Defendants:

Brock, Clay & Calhoun, P.C.

Carol Callaway

E. Linwood Gunn, IV

Appellees/Plaintiffs:

Kathleen Chapman

Terry Jackson

Paul Mason

Debra Ann Power

Selman v. Cobb County School District, No. 05-10341-II

Jeffrey Michael Selman

Jeff Silver

Counsel for Appellees/Plaintiffs:

American Civil Liberties Union Foundation

American Civil Liberties Union Foundation of Georgia, Inc.

Bondurant, Mixson & Elmore, LLP

David G. H. Brackett

Jeffrey O. Bramlett

Margaret F. Garrett

Emily Hammond Meazell

Gerald R. Weber

Proposed Intervenors:

Allen Hardage

Larry Taylor

Counsel for Proposed Intervenors:

Alliance Defense Fund

Hollberg & Weaver, LLP

Kevin Thomas McMurry

Kevin H. Theriot

Selman v. Cobb County School District, No. 05-10341-II

George M. Weaver

Amici Curiae:

Biologists and other Scientists

Colorado Citizens for Science

Honorable J. Foy Guin, Jr.

Kansas Citizens for Science

Michigan Citizens for Science

Nebraska Religious Coalition for Science Education

New Mexico Academy of Science

New Mexico Coalition for Excellence in Science and Math Education

New Mexicans for Science and Reason

Parents for Truth in Education

Texas Citizens for Science

Counsel for Amici:

Seth L. Cooper
Center for Science & Culture
Discovery Institute

David DeWolf

Lynn Gitlin Fant

Hollberg & Weaver, LLP

C- 3

Selman v. Cobb County School District, No. 05-10341-II

William Johnson, Esq.

Rogers & Watkins, LLP

Marjorie Rogers

George M. Weaver

Trial Judge:

Honorable Clarence Cooper

TABLE OF CONTENTS

TABLE OF AUTHORITIES

Scientific Authorities

STATEMENT OF ISSUES

1. Are there legitimate scientific issues as to whether life arose and developed by means of chemical and biological evolution?
2. May the legitimate scientific issues concerning chemical and biological evolution be discussed in the public school classroom without endorsing a religion?

INTRODUCTION

Amici curiae are scientists and include a number of biologists. Most of them live in the Eleventh Circuit's jurisdiction. Each of the individual signatories to the brief has earned a science-related doctoral degree. Amici include university professors, research scientists and scientists in private industry. All amici question biological or neo-Darwinian evolutionary theory (the modern Darwinian theory of evolution) from a scientific perspective, as well as evolutionary accounts of the chemical origin of the first life on Earth. That is to say, amici are scientists who are skeptical of the ability of random mutations and natural selection to account for the origin and complexity of life.

INTEREST OF AMICI CURIAE

Amici are professional scientists who seek to inform the Court that there is a live and growing scientific controversy surrounding neo-Darwinian theory. This controversy, which is implicated in this case, is the subject of serious academic debate. Amici also seek to highlight the scientific controversy over whether chemical evolutionary theory can adequately explain the origin of the first life on Earth. Finally, Amici assert that the science education necessary to equip students for the 21st Century should not censor relevant scientific information about important scientific controversies (such as neo-Darwinian and chemical evolutionary theories), but should fully inform students about such scientific debates.

COMPLETE LIST OF AMICI CURIAE

Biologists

Raymond G. Bohlin, Ph.D. Molecular and Cell Biology (University of Texas at Dallas);

Yvonne Boldt, Ph.D. Microbiology (University of Minnesota);

William S. Harris, Ph.D. Nutritional Biochemistry (University of Minnesota), Professor of Medicine, University of Missouri-Kansas City School of Medicine;

Cornelius Hunter, Ph.D. Biophysics and Computational Biology (Illinois University);

Dean Kenyon, Ph.D. Biophysics (Stanford University), Professor Emeritus of Biology, San Francisco State University;

Scott Minnich, Ph.D. Microbiology (Iowa State University), Associate Professor of Microbiology, University of Idaho;

Ralph Seelke, Ph.D. Microbiology (University of Minnesota); Professor of Microbiology, University of Wisconsin-Superior.

Chris Williams, Ph.D. Biochemistry (The Ohio State University);

Other Scientists

Gary L. Achtemeier, Ph.D. Meteorology (Florida State University);

Changhyuk An, Ph.D. Physics (University of Tennessee);

Eugene C. Ashby, Ph.D. Chemistry (Notre Dame University), Emeritus Regents Professor and Distinguished Professor, School of Chemistry and Biochemistry, Georgia Institute of Technology;

Phillip Bishop, Ed.D. Exercise Physiology (University of Georgia), Professor of Kinesiology, University of Alabama;

John H. Bordelon, Ph.D. Electrical Engineering (Georgia Institute of Technology), Senior Research Engineer, School of Electrical & Computer Engineering, Georgia Institute of Technology;

Noel Ricky Byrn, Ph. D. Nuclear Engineering (Georgia Institute of Technology);

Nancy Bryson, Ph.D. Chemistry (University of South Carolina), Assistant Professor of Chemistry, Kennesaw State University;

A. Eugene Carden, Ph.D. Metallurgy (University of Connecticut), Professor Emeritus of Engineering Mechanics, University of Alabama;

Russell W. Carlson, Ph.D. Biochemistry (University of Colorado, Boulder), Professor of Biochemistry & Molecular Biology, Technical Director of the Complex Carbohydrate Research Center, University of Georgia;

Leon L. Combs, Ph.D. Chemical Physics (Louisiana State University), Professor and Chair, Department of Chemistry and Biochemistry, Kennesaw State University;

Michael Covington, Ph.D. Linguistics (Yale University), Associate Director, Artificial Intelligence Center, University of Georgia;

Malcolm A. Cutchins, Ph.D. Engineering Mechanics (Virginia Polytechnic Institute and State University), Emeritus Professor of Aerospace Engineering, Auburn University;

Cham E. Dallas, Ph.D. Toxicology (University of Texas, Austin), Professor and Director, CDC Center for Mass Destruction Defense, University of Georgia and Medical College of Georgia;

S. Todd Deal, Ph. D. BioOrganic Chemistry (The Ohio State University), Professor of Chemistry, Georgia Southern University;

Keith S. Delaplane, Ph.D. Entomology (Louisiana State University), Professor of Entomology, University of Georgia;

Allison J. Dobson, Ph.D. Physical Chemistry (The Ohio State University), Associate Professor of Chemistry, Georgia Southern University;

John M Ford, Ph.D. Physics (Virginia Polytechnic Institute and State University);

Christian Heiss, Ph.D. Chemistry (University of Georgia);

Dewey H. Hodges, Ph.D. Aeronautical & Astronautical Engineering (Stanford University), Professor, Aerospace Engineering, Georgia Institute of Technology;

Timothy Hoover, Ph.D. Biochemistry (University of Wisconsin), Associate Professor and Associate Head of Microbiology, University of Georgia;

Richard J. Kinch, Ph.D. Electrical Engineering and Computer Science (Cornell University);

Terrie L. Lampe, Ph.D. Chemistry (Wayne State University), Professor of Chemistry, Georgia Perimeter College;

Joseph M. Lary, Ph.D. Biology (University of Alabama);

George Lebo, Ph.D. Physics (University of Florida), Emeritus Associate Professor Astronomy, University of Florida;

Roger J. Lien, Ph.D. Physiology (North Carolina State University), Associate Professor, Poultry Science Department, Auburn University;

Emerson Thomas McMullen, Ph.D. History & Philosophy of Science (Indiana University), Associate Professor of History, Georgia Southern University;

Henry F. Schaefer, Ph.D. Chemical Physics (Stanford University), Graham Perdue Professor of Chemistry and Director of the Center for Computational Chemistry, University of Georgia;

Norman E. Schmidt, Ph.D. Chemistry (University of South Carolina), Professor of Chemistry, Georgia Southern University;

Robert B. Sheldon, Ph.D. Physics (University of Maryland, College Park);

Michael A. Skinner, M.D. (Rush College of Medicine), Associate Professor of Surgery, Duke University;

William C. Small, M.D., Ph.D., Physical Chemistry (Emory University), Associate Professor of Radiology, Emory University;

Darwin W. Smith, Ph.D., Chemistry (California Institute of Technology), Emeritus Professor of Chemistry, University of Georgia;

Daniel W. Tedder, Ph.D. Chemical Engineering (University of Wisconsin), Associate Professor, School of Chemical and Biomolecular Engineering, Georgia Institute of Technology;

Charles B. Thaxton, Ph.D. Physical Chemistry (Iowa State University), co-author, The Mystery of Life's Origin: Reassessing Current Theories (1984);

James A. Tumlin, M.D. (Emory University), Associate Professor of Medicine, Emory University;

William E. Wade, Pharm.D., (University of Georgia), Professor of Pharmacy, College of Pharmacy, University of Georgia;

A. Bruce Webster, Ph.D., Department of Animal and Poultry Science (University of Guelph, Canada);

Robert Wentworth, Ph.D. Toxicology (University of Georgia), Health and Safety Coordinator, Office of Human Resources, University of Georgia;

Mark G. White, Ph.D. Chemical Engineering (Rice University), Professor of Chemical and Biomolecular Engineering, Georgia Institute of Technology;

John W. Worley, Ph.D. Agricultural Engineering (Virginia Polytechnic Institute and State University), Associate Professor, Department of Biological and Agricultural Engineering, University of Georgia.

SUMMARY OF ARGUMENT

Amici scientists wish to bring to the Court's attention the current debate within scientific disciplines over whether chemical and biological (i.e., neo-Darwinian) evolution can adequately account for the origin of life and the development of life into its current forms. This debate is scientific and not religious in nature.

In order for public school students to receive an adequate scientific education, they should be acquainted with the debate over chemical and neo-Darwinian evolution. This debate can be discussed without practicing religion or even referring to religion. Amici contend that the sticker placed by the Cobb County School Board sticker in certain science textbooks, to the extent it encourages students to think critically and grapple with the scientific debate, is not unconstitutional. Importantly, the sticker does not even endorse or mention religion.

ARGUMENT AND CITATIONS OF AUTHORITY

Scientific discoveries of the last few decades have led to greater skepticism over the ability of the mechanisms of biological or neo-Darwinian evolutionary theory to account for the complexity of life we see today. Amici represent a sampling of the growing number of scientists who are skeptical of neo-Darwinism's claim that the undirected mechanisms of natural selection and random genetic variations can account for the complexity of life. Amici also represent a number of scientists who are skeptical of chemical evolutionary theory's ability to account for the origin of life.

As the district court recognized, that there are scientists who continue to raise scientific challenges to neo-Darwinian and chemical evolutionary theories.[1] Amici are doctoral scientists who are skeptical of neo-Darwinian theory and chemical evolutionary theory on scientific grounds. Neo-Darwinian theory is being re-examined by scientists in light of new scientific discoveries. Scientific discoveries of the past few years and the increasing body of scientific knowledge

[1](R4-98-33) ("there are some scientists who have questions regarding certain aspects of evolutionary theory").

available today makes the claims of neo-Darwinian theory far less tenable than in the early part of the 20th Century. One biochemist has gone so far as to describe neo-Darwinian theory as "a theory in crisis."[2] An increasing number of scientific publications directly challenge neo-Darwinian theory, or key aspects of it.[3] Recent discoveries have also led to greater challenges for traditional chemical evolutionary scenarios for the origin of the first life from non-life.

Neo-Darwinian theory presently remains the dominant theory of origins in the scientific community, but serious debate now exists about its sufficiency.

---

[2]Michael J. Denton, Evolution: A Theory in Crisis (1986).

[3]*See, e.g.*, Michael J. Behe and David W. Snoke, *Simulating Evolution by Gene Duplication of Protein Features that Require Multiple Amino Acid Residues*, 13 Protein Science 2651 (October, 2004); Michael J. Behe, *Irreducible Complexity: Obstacle to Darwinian Evolution*, in William A. Dembski and Michael Ruse, eds., Debating Design: From Darwin to DNA 352 (2004); Michael J. Behe, *Self-Organization and Irreducibly Complex Systems: A Response to Shanks and Joplin,* 67 Philosophy of Science 155-62 (March, 2000); Michael J. Behe, Darwin's Black Box: The Biochemical Challenge to Evolution (1996); William A. Dembski, No Free Lunch: Why Specified Complexity Cannot be Purchased Without Intelligence (2002); Michael J. Denton, Nature's Destiny (1998); Michael J. Denton, Evolution: A Theory in Crisis; Stephen C. Meyer, *The Origin of Biological Information and the Higher Taxonomic Categories*, 117 [no. 2] Proceedings of the Biological Society of Washington 213-239 (2004); Stephen C. Meyer, Marcus Ross, Paul Nelson, and Paul Chien, *The Cambrian Explosion: Biology's Big Bang,* in John Angus Campbell and Stephen C. Meyer, eds., Darwin, Design and Public Education 323 (2003); Scott A. Minnich and Stephen C. Meyer, *Genetic Analysis of Coordinate Flagellar And Type III Regulatory Circuits in Pathogenic Bacteria*, Second International Conference on Design & Nature (2004); Jeffrey H. Schwartz, Sudden Origins: Fossils, Genes and the Emergence of Species (1999).

Although amici represent a minority position within the scientific community, dissenting viewpoints have always been an integral part of the scientific process. Scientists debate over how best to interpret data. When such debates are raging, students need to know about them.

In addition to amici and other scientists who are skeptical of neo-Darwinian theory, there are many scientists who still accept the theory but acknowledge some of its difficulties. Many such scientists have pointed to scientific problems surrounding aspects of neo-Darwinian theory.[4]

There are two main parts of neo-Darwinian evolutionary theory: universal common descent and the power of natural selection. Scientific publications highlight neo-Darwinian theory's problems related to pattern; i.e. the large-scale geometry of biological history.[5] Questions remain as to how organisms are related

---

[4]*See, e.g.*, selected essays in Gerd B. Müller and Stuart A. Newman, eds., Origination of Organismal Form: Beyond the Gene in Developmental and Evolutionary Biology (2003); James W. Valentine, On the Origin of Phyla 189-194 (2004).

[5]*See, e.g.*, Michael S. Y. Lee, *Molecular Clock Calibrations and Metazoan Divergence Dates*, 49 Journal of Molecular Evolution 385 (1999); Michael S. Y. Lee, *Molecular Phylogenies Become Functional*, 14 Trends in Ecology and Evolution 177-178 (1999); Simon Conway Morris, *Evolution: Bringing Molecules into the Fold*, 100 Cell 1 (2000); Simon Conway Morris, *The Cambrian 'Explosion' of Metazoans*, in Origination of Organismal Form 13 (2003); Simon Conway Morris, *The Question of Metazoan Monophyly and the Fossil Record*, 21 Progress in Molecular and Subcellular Biology 1 (2003); Simon Conway Morris, *Cambrian 'Explosion' of Metazoans and Molecular Biology: Would Darwin be*

to one another and how we can detect such relationships. An increasing number of scientists have raised questions about whether there is sufficient evidence for universal common descent.

Other scientific publications underscore Darwinian theory's difficulties concerning process; i.e., the mechanisms of evolution.[6] Questions persist as to

---

Satisfied?, 47 (7-8) International Journal of Developmental Biology 505 (2003); James W. Valentine, & D. Jablonski, *Morphological and developmental macroevolution: a paleontological perspective*, 47 International Journal of Developmental Biology 517-522 (2003); P. Willmer, *Convergence and Homoplasy in the Evolution of Organismal Form*, in Origination of Organismal Form 33-50 (2003); P. Willmer, Invertebrate Relationships: Patterns in Animal Evolution (1990); Carl Woese, *The Universal Ancestor*, 95 Proceedings of the National Academy of Sciences USA 6854-6859 (1998).

[6]*See, e.g.*, H. Becker & W. Lonnig, *Transposons: Eukaryotic*, in 18 Nature Encycolpedia of Life Sciences, 529 (2001); Michael J. Behe and David W. Snoke, *Simulating Evolution by Gene Duplication of Protein Features that Require Multiple Amino Acid Residues*, 13 Protein Science 2651 (October, 2004); R. L. Carroll, *Towards a New Evolutionary Synthesis*, 15 Trends in Ecology and Evolution 27-32 (2000); D. H. Erwin, *Early Introduction of Major Morphological Innovations*, 38 Acta Palaeontologica Polonica 281 (1994); S.F. Gilbert, et al., *Resynthesizing Evolutionary and Developmental Biology*, 173 Developmental Biology 357 (1996); B. C. Goodwin, *What are the Causes of Morphogenesis?* 3 BioEssays 32-36 (1985); W. E. Lonnig & H. Saedler, *Chromosome Rearrangements and Transposable Elements*, 36 Annual Review of Genetics 389 (2002); Simon Conway Morris, *Evolution: Bringing Molecules into the Fold*, 100 Cell 1 (2000); Simon Conway Morris, *Cambrian "Explosion" of Metazoans and Molecular Biology: Would Darwin Be Satisfied?*, 47 [7-8] International Journal of Developmental Biology 505 (2003); Olivier Rieppel, *Turtles as Hopeful Monsters*, 23 BioEssays 987-91 (2001); N. H. Shubin & C. R. Marshall, *Fossils, Genes and the Origin of Novelty*, in Deep Time 324 (2000); B. M. Stadler, et. al, *The Topology of the Possible: Formal Spaces Underlying Patterns of Evolutionary Change*, 213 Journal of Theoretical Biology 241 (2001); K. S. Thomson,

whether microevolutionary processes can be extrapolated to prove macroevolutionary change. Still other scientific publications call into question the ability of neo-Darwinian mechanisms to generate novel genetic information, novel organs, structures and body plans.

In addition, many scientific publications have questioned whether chemical evolutionary theory can explain the origin of the first life from non-living chemicals (the "origin-of-life" problem).[7]

Macroevolution: The Morphological Problem, 32 American Zoologist 106 (1992); James W. Valentine, On the Origin of Phyla 189-94 (2004); G. P. Wagner & P.F. Stadler, *Quasi-Independence, Homology and the Unity-C of Type: A Topological Theory of Characters*, 220 Journal of Theoretical Biology 505 (2003); G. Webster & B. Goodwin, Form and Transformation: Generative and Relational Principles in Biology (1996).

For discussion of many of the above references, see Stephen C. Meyer, Stephen C. Meyer, *The Origin of Biological Information and the Higher Taxonomic Categories*, 117 [no. 2] Proceedings of the Biological Society of Washington 213 (2004).

[7]*See, e.g.*, Simon Conway Morris, Life's Solution: Inevitable Humans in a Lonely Universe 22-43, and esp. 44-68 (2003); Paul Davies, The Fifth Miracle: The Search for the Origin and Meaning of Life (2000); Leslie E. Orgel, *The Origin of Life—A Review of Facts and Speculations*, 23 Trends in Biochemical Science 491-95 (1998); Antonio Lazcano & Stanley Miller, *The Origin and Early Evolution of Life: Prebiotic Chemistry, the Pre-RNA World, and Time*, 85 Cell 793 (1996); Hubert Yockey, Information Theory and Molecular Biology, esp. 259-293(1992); Robert Shapiro, Origins: A Skeptic's Guide to the Creation of Life on Earth, esp. 132-154 (1986); Robert Shapiro, *Prebiotic Ribose Synthesis: A Critical Analysis*, 18 Origins of Life & Evolution Biosphere 71, 71-85 (1988).

Also see Walter Bradley, *Information, Entropy and the Origin of Life*, in Debating Design 331 (2004); Stephen C. Meyer, *DNA and the Origin of Life: Information, Specification and Explanation*, in Darwinism, Design and Public

Amici emphasize that standard high school and college biology textbooks routinely ignore scientific data challenging neo-Darwinian and chemical evolutionary theories, as well as scientific data that merely point to widely-acknowledged scientific problems confronting those theories.

Furthermore, many textbooks contain alleged evidences for neo-Darwinian theory that have long been discredited by scientists, including neo-Darwinists.[8] Amici assert that school boards should be able to take reasonable steps to ensure that students are fully-informed about the scientific controversy surrounding Darwin's theory and that their curriculum is free from factual errors, including

---

Education 223 (2003); Charles B. Thaxton, et al., The Mystery of Lifes's Origin: Reassessing Current Theories (1984).

[8]*See, e.g.*, Jerry Coyne, *Not Black and White*, 396 Nature 35-36 (1998); Stephen Jay Gould, *Abscheulich! (Atrocious!)*, Natural History 42-49 (March, 2000); Judith Hooper, Of Moths & Men: An Evolutionary Tale: The Untold Story of Science and the Peppered Moth (2002); Craig Millar & David Lambert, *Industrial Melanism–A Classic Example of Another Kind?*, 49 BioScience 1021-23 (1999); Elizabeth Pennisi, *Haeckel's Embryos: Fraud Rediscovered*, 277 Science 1435 (1997); Michael Richardson, et al., *There is No Highly Conserved Embryonic Stage in the Vertebrates: Implications for Current Theories of Evolution and Development*, 196 Anatomy & Embryology 91 (1997); Theodore D. Sargent, Craig D. Millar & David Lambert, *The 'Classical' Explanation of Industrial Melanism: Assessing the Evidence*, 30 Evolutionary Biology 299 (1998); Jonathan Wells, *Haeckel's Embryos & Evolution: Setting the Record Straight*, American Biology Teacher 345-49 (May, 1999); Jonathan Wells, *Second Thoughts About Peppered Moths*, 11 Scientist 11 (May 24, 1999).

Also see Jonathan Wells, Icons of Evolution: Science or Myth?: Why Much of What We Teach About Evolution is Wrong (2000).

those that overstate the case for neo-Darwinian theory and chemical evolutionary theory.

In some instances, it is likely that metaphysical preferences and presuppositions of some scientists have prevented students from learning about scientific challenges to neo-Darwinian and chemical evolutionary theories or prevented the correction of textbook errors that overstate the case for neo-Darwinian and chemical evolutionary theories.

The lack of public science classroom coverage given to the growing scientific controversy surrounding neo-Darwinian evolutionary theory and frequent inclusion of erroneous information about the subject in textbooks (without any corrective counter-balances) present a dilemma for many school board members, administrators and educators who wish to teach neo-Darwinian and chemical evolutionary theories—but also wish to do so in the fairest and most accurate manner possible.

Amici fully support the United States Supreme Court's recognition of the power of states and local school boards to permit teachers and students to discuss scientific critiques of prevailing scientific theories. Regardless of this Court's ultimate determination of the constitutional status of the textbook sticker at issue[9]—and in light of the controversy over neo-Darwinism, the controversy over the chemical origin of life, and importance of critical thinking skills as a part of good science education—Amici urge the Court to be mindful of the importance of academic freedom in public school classrooms and the essential role of dissenting scientific viewpoints to the scientific enterprise.

CONCLUSION

For the foregoing reasons, amici curiae urge the Court to reverse the decision of the district court and uphold the constitutionality of the textbook sticker.

Respectfully submitted,

____

GEORGE M. WEAVER* 743150
Hollberg & Weaver, LLP
Attorneys for Amici Curiae

* Counsel of Record

2941 Piedmont Road, N.E. – Suite C
Atlanta, GA 30305

[9]The sticker states:

> This textbook contains material on evolution. Evolution is a theory, not a fact, regarding the origin of living things. This material should be approached with an open mind, studied carefully, and critically considered.

(R4-98-8).

Notes

Introduction

1. See "Kurt Vonnegut on Darwinism and Intelligent Design," http://www.evolutionnews.org/2006/01/kurt_vonnegut_on_darwinism_and.html (accessed September 11, 2006).
2. For examples of Darwinian conservatives, see Charles Krauthammer, "Phony Theory, False Conflict: 'Intelligent Design' Foolishly Pits Evolution Against Faith," *The Washington Post*, November 18, 2005, p. A23; George Will, "Grand Old Spenders," *The Washington Post*, November 17, 2005, http://www.washingtonpost.com/wp-dyn/content/article/2005/11/16/AR2005111601883.html (accessed September 5, 2006); John Derbyshire, "Intelligent Design Roundup," http://www.nationalreview.com/thecorner/05_02_06_corner-archive.asp - 055599 (accessed August 8, 2006); James Q. Wilson, "Faith in Theory: Why 'intelligent design' simply isn't science," *The Wall Street Journal*, December 26, 2005, available at http://www.opinionjournal.com/editorial/feature.html?id=110007726 (accessed August 23, 2006); and Larry Arnhart, *Darwinian Natural Right: The Biological Ethics of Human Nature* (New York: State University of New York Press, 1998) and *Darwinian Conservatism* (Charlottesville, VA: Imprint Academic, 2005).
3. Will, "Grand Old Spenders."
4. Krauthammer, "Phony Theory"; Wilson, "Faith in Theory."
5. James Q. Wilson, *The Moral Sense* (New York: The Free Press, 1993), especially chapters 6–8.
6. John O. McGinnis, "The Origin of Conservatism," *National Review*, December 22, 1997, p. 31.
7. Arnhart, *Darwinian Conservatism*, p. 1.
8. Ibid., p. 8.
9. For example: "Darwin employed this idea of spontaneously evolving order to explain the evolution of complex structures and processes in living things through random heritable variation with selective retention by natural selection." [Ibid., p. 17]; and "Darwin's aim was to show how this moral sense could have arisen in human nature as an evolutionary product of natural selection." [Ibid., p. 37]
10. See, for example, Arnhart's discussion of the biological basis of differences be-

tween the sexes, ibid., pp. 51–58.

11. "Human culture is a spontaneous order, because the customs of social life develop over history through an evolutionary process of random variation and selective retention to create complex cultural order that has not been intentionally designed." Ibid., p. 25.
12. For a good discussion of the various meanings of "evolution," see Stephen C. Meyer and Michael Newton Keas, "The Meanings of Evolution," in John Angus Campbell and Stephen C. Meyer, editors, *Darwinism, Design, and Public Education* (East Lansing, MI: Michigan State University Press, 2003), pp. 135–156.
13. Often cited by evolutionists today, this phrase originated with neo-Darwinist Theodosius Dobzhansky in his article "Nothing in Biology Makes Sense Except in the Light of Evolution," *The American Biology Teacher*, vol. 35 (1973), pp. 125–129.
14. Philip S. Skell, "Why Do We Invoke Darwin? Evolutionary theory contributes little to experimental biology," *The Scientist*, August 28, 2005, issue 16, p. 10. To make this point, Skell quotes A.S. Wilkins, editor of the journal *BioEssays*, who has written: "Evolution would appear to be the indispensable unifying idea and, at the same time, a highly superfluous one."
15. George Gaylord Simpson, *The Meaning of Evolution: A Study of the History of Life and of Its Significance for Man*, revised edition (New Haven: Yale University Press, 1967), p. 345.
16. Charles Darwin, *The Variation of Animals and Plants under Domestication*, second edition (New York: D. Appleton & Co., 1883), vol. II, pp. 428–429.
17. Salvador Luria, Stephen Jay Gould, Sam Singer, *A View of Life* (Menlo Park, CA: Benjamin/Cummings Publishing Company, 1981), pp. 586–587. More generally, see the book's subsection titled "The Radical Philosophy of Darwin's Theory," pp. 584–587.
18. Kenneth R. Miller and Joseph Levine, *Biology*, fourth edition (Upper Saddle River, NJ: 1998), p. 658.
19. Douglas Futuyma, *Evolutionary Biology*, third edition (Sunderland, MA: Sinauer Associates, 1998), p. 5.
20. Ibid., p. 8, emphasis in the original.
21. William K. Purves, David Sadava, Gordon H. Orians, and H. Craig Heller, Life: *The Science of Biology*, sixth edition (Sunderland, MA: Sinauer Associates, 2001; Salt Lake City: W. H. Freeman and Co., 2001), p. 3. For nearly the same wording, see William K. Purves, Gordon H. Orians, H. Craig Heller, and David Sadava, Life: *The Science of Biology*, fifth edition (Sunderland, MA: Sinauer Associates, 1998; Salt Lake City: W. H. Freeman and Co., 1998), p. 10.] ." Textbooks that put forward such metaphysically-loaded statements typically also offer the obligatory caveats that of course Darwin's theory and religion are not necessarily incompatible. But these caveats are hard to take seriously. Asserting that the development of life is "purposeless," that supernatural explanations are "superfluous," and that embracing Darwin's theory "require[s] believing in philosophical materialism" are very peculiar ways of demonstrating to students that Darwinism and theism are harmonious.

22. "NABT Unveils New Statement on Teaching Evolution," *The American Biology Teacher* (January 1996), p. 61.
23. "Eugenie Scott's Reply to the Open Letter," http://fp.bio.utk.edu/darwin/Open%20letter/scott%20reply.htm (accessed June 17, 2002).
24. See Larry Witham, "Evolutionists Split over Biology Lessons" (February 24, 1998), p. A2; "Evolution Statement Altered," *The Christian Century* (Nov. 12, 1997), p. 1029.
25. Eugenie Scott, "NABT Statement on Evolution Evolves," http://www.ncseweb.org/resources/articles/8954_nabt_statement_on_evolution_ev_5_21_1998.asp (accessed July 16, 2005).
26. In 2004, the NABT statement evolved yet again. Apparently the "no specific direction or goal" language continued to pose a public relations problem, and so it was toned down. According to the 2004 version, "natural selection has no discernible direction or goal." "NABT's Statement on Teaching Evolution, Supporting Material," http://www.nabt.org/sub/position_statements/evolution_supporting-material.asp (accessed July 16, 2005), emphasis added.
27. Eugenie Scott, quoted in an article in *More* magazine, July–August 2005.
28. "Open Letter to the National Association of Biology Teachers, to the National Center for Science Education, and to the American Association for the Advancement of the Sciences," http://fp.bio.utk.edu/darwin/Open%20letter/openletter.html (accessed June 17, 2002).
29. *Kansas Science Education Standards*, as approved by the Kansas State Board of Education on Nov. 8, 2005, p. 75. Available from the Kansas State Department of Education at http://www.ksde.org/outcomes/sciencestd.pdf (accessed August 8, 2006).
30. Letter from Nobel Laureates to Kansas State Board of Education, September 9, 2005. The letter was sent out under the auspices of the Elie Wiesel Foundation, http://www.eliewieselfoundation.org/. A copy of the letter is posted at http://media.ljworld.com/pdf/2005/09/15/nobel_letter.pdf (accessed August 8, 2006).
31. Ibid. Emphasis added.

I. Does Darwinism Support Traditional Morality?

32. Michael Ruse and E.O. Wilson, "The Evolution of Ethics" in James Huchingson, *Religion and the Natural Sciences: The Range of Engagement* (New York: Harcourt Brace Jovanovich, 1993), p. 210.
33. Thomas Huxley, "Evolution and Ethics,"in Thomas H. Huxley, *Evolution and Ethics and Other Essays* (New York: D. Appleton and Co., 1925), p. 81–82.
34. Ibid., p. 83.
35. Wilson, *The Moral Sense*. For example, "[e]volution has selected for attachment behavior in all species that nurture their young after birth.... Theories of natural selection and inclusive fitness explain why caring for one's own young has been adaptive—that is, useful—for the human species. People who care for their young leave more young behind than those who do not; to the extent parental care is under genetic control, caring parents reproduce their genes in the next generation

at higher rates than do uncaring ones." Ibid., p. 126.

36. Arnhart, *Darwinian Conservatism*, p. 35.
37. Ibid., p. 23.
38. Charles Darwin, *The Descent of Man, and Selection in Relation to Sex* (Princeton: Princeton University Press, 1981), vol. I, p. 106. This is a reprint of the the first edition, and hereafter cited as *Descent* (1871).
39. Ibid., p. 82.
40. Ibid., vol. II, p. 394.
41. Ibid., vol. I, pp. 83–84; vol. II, pp. 363–65.
42. Ibid., vol. I, pp. 76–77.
43. Ibid., p. 103.
44. Ibid., p. 104.
45. Ibid., p. 73.
46. For an account of Darwin's kindly and sympathetic nature in private life, see Francis Darwin, editor, *The Autobiography of Charles Darwin and Selected Letters* (New York: Dover, 1958), especially pp. 303–04.
47. "Universal human nature," appears to be Arnhart's term of choice; see in particular p. 23 of *Darwinian Conservatism* ("the only escape from... cultural relativism is to argue that there is a universal human nature."). But on p. 8 of the same book, Arnhart uses the term "unchanging human nature" to approvingly summarize Thomas Sowell's account of the "realist vision," and on p. 130 he talks of "an enduring human nature" ("Darwinian science presents an enduring human nature that cannot be treated as mere matter to be shaped by social planners.").
48. For Arnhart's effort to grapple with the question of the permanency of Darwinian human nature in the area of sex differences, see *Darwinian Conservatism*, pp. 54–55.
49. Arnhart, *Darwinian Natural Right*, p. 229.
50. For a similar critique of Arnhart's view on this point, see J. Budziszewski, "Accept No Imitations: The Rivalry of Naturalism and Natural Law," in William A. Dembski, editor, *Uncommon Dissent: Intellectuals Who Find Darwinism Unconvincing* (Wilmington, DE: ISI Books, 2004), pp. 109–110.
51. Ibid., p. 110.
52. Arnhart, *Darwinian Natural Right*, p. 229.

II. Does Darwinism Support the Traditional Family?

53. Malcolm Potts and Roger Short, *Ever Since Adam and Eve: The evolution of human sexuality* (Cambridge, United Kingdom: Cambridge University Press, 1999), p. 332.
54. Arnhart, *Darwinian Conservatism*, pp. 10–11. On p. 46, Arnhart similarly declares that "Darwinian science supports the conservative position by showing how marriage, family life, and sex differences conform to the biological nature of human beings as shaped by evolutionary history."
55. Ibid., p. 48.

56. Ibid.
57. Charles Darwin, *The Descent of Man and Selection in Relation to Sex*, second edition (London: John Murray, 1882), p. 590. This edition of *The Descent of Man* is hereafter cited as *Descent* (1882).
58. Arnhart, *Darwinian Conservatism*, p. 48.
59. Darwin, *Descent* (1871), vol. II, p. 362.
60. Darwin, *Descent* (1882), p. 591.
61. Ibid., p. 591. Darwin similarly stated a couple of pages earlier: "Men and women, like many of the lower animals, might formerly have entered into strict though temporary unions for each birth. *Descent* (1882), p. 589; also see *Descent* (1871), vol. II, pp. 360, 362. In the second edition of *The Descent of Man*, Darwin did rewrite a sentence from the first edition that most explicitly revealed his heterodox definition of man's original "monogamy": "Turning to primeval times when men had only doubtfully attained the rank of manhood, they would probably have lived, as already stated, either as polygamists or temporarily as monogamists." Ibid., p. 367, emphasis added. In the second edition, Darwin downplayed the reference to polygamy and removed the explicit admission of the *temporary* nature of early monogamy. The revised sentence read: "Judging from the analogy of the lower animals, he [man] would then either live with a single female, or be a polygamist." *Descent* (1882), p. 594.
62. See Edward Westermarck, *The History of Human Marriage* (London: Macmillan, 1921), vol. I, II, III.
63. Following is a selection of passages in which Westermarck cites the importance of natural selection in the develop of human sexuality: "If we ask why in certain animal species male and female remain together not only during the pairing season but till after the birth of the offspring, I think that there can be no doubt as regards the true answer. They are induced to do so by an instinct which has been acquired through the process of natural selection beause it has a tendency to preserve the next generation and thereby the species." [Westermark, *The History of Human Marriage*, vol. I, p. 35]; "Marriage... seems to be based upon a primeval habit. We have found reasons to believe that even in primitive times it was the habit for a man and a woman, or several women, to remain together till after the birth of the offspring, and that they were induced to do so by an instinct which had been acquired through natural selection because the offpsring were in need of both maternal and paternal care." [Ibid., vol. III, p. 365]; Discussing the incest taboo, Westermarck made a general comment that "any satisfactory explanation of the normal characteristics of the sexual instinct, which is of such immense importance for the existence of the species, must be sought for in their specific usefulness. We may assume that in this, *as in other cases*, natural selection has operated, and by eliminating destructive tendencies and preserving useful variations has moulded the sexual instinct so as to meet the requirements of the species." [Edward Westermarck, *The Future of Marriage in Western Civilisation* (New York: Macmillan, 1936), pp. 258–259, emphasis added.] More generally, for the influence of Darwin on Westermarck, see Edward Westermarck, *Memories of My Life*, translated from the Swedish by Anna Barwell, (New York: The Macaulay Co., 1929), pp. 67–68, 77–79, 88–89, 91–92. Westermarck stated that "Darwin's book

on the *Descent of Man*... proved of the greatest importance for my future work" [Ibid., p. 67] and that he "found the theory of natural selection, especially in its application to instinct, to be of enormous importance for the solution of many of the problems with which I was occupied." [Ibid., p. 77] At the same time, he criticized Darwin's theory of sexual selection, and his Lamarckian explanation for the origin of new variations, preferring the mutation theory that later formed the basis of what became known as "neo-Darwinism." Darwin's greatest accomplishment according to Westermarck was the banishment of design from explanations of biology: "Darwin discovered one factor which must have influenced evolution in a greater or less degree; and his discovery was undeniably of such a nature that it surpassed all later biological discoveries in its effect upon our general view of life. He gave an explanation of the appearance of purpose in organic life without calling in the aid of the hypothesis of a providence that has created the world according to a certain plan drawn up after a human pattern." [Ibid., pp. 78–79] For more on the relationship between Westermarck and Darwin's theory, see Timothy Stroup, *Westermarck's Ethics* (Åbo: Research Institute of the Åbo Akademi Foundation, 1982), pp. 35–38, 66–68, 129–131; Timothy Stroup, "Introduction," in Timothy Stroup, editor, *Edward Westermarck: Essays on His Life and Works* (Helsinki: The Philosophical Society of Finland, 1982), pp. xiii–xiv; Georg Henrik Von Wright, "The Origin and Development of Westermarck's Moral Philosophy," in Stroup, *Edward Westermarck: Essays*, pp. 29–30, 39–41.

64. See discussion in Westermarck, *The Future of Marriage*, pp. 151–171.
65. According to Westermarck, "the emotional origin of moral judgments consistently leads to a denial of the objective validity ascribed to them both by common sense and by normative theories of ethics." [Westermarck, *Ethical Relativity* (New York: Harcourt, Brace and Company, 1932), p. xvii.] For Westermarck's development of moral relativism, see not only his *Ethical Relativity*, but *The Origin and Development of the Moral Ideas* (New York: Macmillan, 1906), vol. I and II.
66. For Westermarck's view of homosexuality, see Westermarck, *The Future of Marriage*, pp. 254–255.
67. Ibid., pp. 244–245.
68. Ibid., p. 201.
69. For the best biography of Kinsey, see James Jones, *Alfred C. Kinsey: A Public/Private Life* (New York: W.W. Norton, 1997). For a thorough critique of his research and worldview, see Judith A. Reisman, *Kinsey: Crimes and Consequences*, second revised edition (Crestwood, KY: Institute for Media Education, 2000).
70. Alfred C. Kinsey, Wardell B. Pomeroy, Clyde E. Martin, *Sexual Behavior in the Human Male* (Philadelphia: W. B. Saunders Co., 1948), pp. 668, 670–674.
71. "The student of human folkways is inclined to see a considerable body of superstition in the origins of all such taboos, even though they may ultimately become religious and moral issues for whole nations and whole races of people." Ibid., p. 669.
72. Ibid., emphasis added.
73. Ibid., p. 222.
74. Ibid., p. 589.

75. Randy Thornhill and Craig T. Palmer, *A Natural History of Rape: Biological Bases of Sexual Coercion* (Cambridge, Mass.: MIT Press, 2000), p. 84.
76. Robert Wright, *The Moral Animal: Evolutionary Psychology and Everyday Life* (New York: Vintage Books, 1995), p. 87.
77. David M. Buss, *Evolutionary Psychology: The New Science of the Mind* (Boston: Allyn and Bacon, 1999), p. 185.
78. Carol M. Ostrom, "Not Meant for Monogamy?" *The Seattle Times*, February 3, 1998, p. D-1.

III. Is Darwinianism Compatible with Free Will?

79. William Provine, abstract for "Evolution: Free will and punishment and meaning in life," talk delivered on Feb. 12, 1998, posted at the Darwin Day Archives, http://eeb.bio.utk.edu/darwin/DarwinDayProvineAddress.htm (accessed August 8, 2006).
80. *Descent* (1871), vol. I, p. 89.
81. Ibid., vol. I, p. 91.
82. Paul Barrett, et. al., *Charles Darwin's Notebooks*, 1836–1844 (New York: Cornell University Press, 1987), 608.
83. Quoted in Frederick Albert Lange, *History of Materialism*, trans. by Ernest Chester Thomas (London: Kegan Paul, Trench, Trübner, & Co. Ltd., 1892), vol. II, p. 312.
84. Cesare Lombroso, *Crime: Its Causes and Remedies*, translated by Henry Horton (Montclair, New Jersey: Patterson Smith, 1968). For a discussion of Lombroso and Social Darwinism see, Mike Hawkins, *Social Darwinism in European and American Thought, 1860–1945* (Cambridge, Great Britain: Cambridge University Press, 1997), pp. 74–80.
85. Ferri, "The Positive School," pp. 137–138.
86. Steven Pinker, professor of psychology at MIT, "Why They Kill Their Newborns," *The New York Times Magazine* (November 2, 1997).
87. Wright, *The Moral Animal*, p. 350.
88. Ibid., p. 37.
89. Ibid.
90. Ibid., p. 88.
91. Provine, "Evolution."
92. David P. Barash, "Dennett and the Darwinizing of Free Will," *Human Nature Review* (March 22, 2003), http://human-nature.com/nibbs/03/dcdennett.html (accessed August 8, 2006).
93. Arnhart, *Darwinian Conservatism*, p. 104.
94. Ibid.
95. Ibid., p. 111.
96. Schwartz is a signer of "A Scientific Dissent from Darwinism," http://www.dissentfromdarwin.org/ (accessed August 8, 2006), and he is a Fellow of the pro-intelligent design International Society for Complexity, Information, and Design,

http://www.iscid.org/jeffrey-schwartz.php (accessed August 8, 2006).

97. Arnhart, *Darwinian Conservatism*, p. 111.
98. Ibid., pp. 104, 106–108.
99. Ibid., p. 110.
100. Ibid.
101. Peter Singer, "Sanctity of Life or Quality of Life?" *Pediatrics*, vol. 72, no. 1 (July 1983), p. 129.
102. Quoted in Arnhart, *Darwinian Conservatism*, p. 126.
103. Ibid., p. 128.
104. Ibid.

IV. Does Darwinism Support Economic Liberty?

105. Quoted in Richard Hoftstadter, *Social Darwinism in American Thought*, revised edition (Boston: The Beacon Press, 1955), p. 51.
106. Thomas Malthus, *An Essay on the Principle of Population* (London, 1798), edited by Ed Stephan, http://www.ac.wwu.edu/~stephan/malthus/malthus.0.html (accessed August 9, 2006).
107. Darwin, *Autobiography*, pp. 42–43.
108. Malthus, *Essay*, chapter 2.
109. Ibid., chapter 7.
110. Francis Bowen, "Malthusianism, Darwinism, and Pessimism," *North American Review*, vol. 129 (1879), p. 452.
111. Ibid., p. 454.
112. Ibid., pp. 455, 456.
113. Robert Bannister, *Social Darwinism: Science and Myth in Anglo-American Social Thought*, with a new preface (Philadelphia: Temple University, 1988), p. 114.
114. See discussion in Tom McKay, "Social Darwinism and Business Thought in the 1920's" (MA Thesis, Western Illinois University, 1973), pp. 47–69.
115. Ludwig von Mises, *Socialism: An Economic and Sociological Analysis*, trans. by J. Kahane (Indianapolis: Liberty Fund, 1981), p. 281.
116. Ibid., p. 285.
117. Ibid., pp. 285–286.
118. F.A. Hayek, "Freedom, Reason and Tradition," *Ethics*, Vol. 68, Issue 4 (July, 1958), p. 232, 233.
119. Arnhart, *Darwinian Conservatism*, p. 18. Also see p. 16, where Arnhart writes: "A spontaneous order is an unintended order. It is a complex order that arises not as the intended outcome of the intelligent design of any mind or group of minds, but as the unintended outcome of many individual actions to satisfy short-term needs. Spontaneous order is design without a designer."
120. See, for example, Robert Hagstrom, *Latticework: The New Investing* (New York: Texere, 2000), especially pp. 55–96; Clay Carr, *Choice, Chance and Organizational Change: Practical Insights from Evolution for Business Leaders and Thinkers* (New

York: Amacom, 1996).

121. William Tucker, "The New Science of Spontaneous Order," *Reason*, January 1996, p. 37.
122. Ibid., p. 38.
123. Ibid.
124. Arnhart, *Darwinian Conservatism*, p. 16.
125. Ibid., p. 18, emphasis added.
126. Ibid., p. 16.
127. Ibid.
128. For another argument for why this is so, see the close reading of Hayek provided by Jay Richards in " Does Hayek's Concept of Spontaneous Order Suggest a World without Purpose?," paper presented to the 81st Annual Conference of the Western Economic Association International, June 29–July 3, 2006, San Diego, California.

V. Does Darwinism Support Limited Government?

129. Woodrow Wilson, *The New Freedom*, with an introduction and notes by William Leuchtenburg (Englewood Cliffs: Prentice-Hall, Inc., 1961), p. 42.
130. See John G. West, "Darwin's Public Policy: Nineteenth Century Science and the Rise of the American Welfare State," in John Marini and Ken Masugi, *The Progressive Revolution in Politics and Political Science: Transforming the American Regime* (Lanham, MD: Rowman and Littlefield, 2005), pp. 261–275; Richard Weikart, *From Darwin to Hitler: Evolutionary Ethics, Eugenics, and Racism in Germany* (New York: Palgrave Macmillan, 2004). Arnhart makes several unfair and inaccurate criticisms of Weikart's work. For Weikart's response, see http://www.csustan.edu/History/Faculty/Weikart/Response-Arnhart.htm (accessed August 26, 2006).
131. Arnhart, *Darwinian Conservatism*, p. 112.
132. For Arnhart's discussion of eugenics, see ibid., pp. 117–122.
133. Mark Haller, *Eugenics: Hereditarian Attitudes in American Thought* (New Brunswick: 1963), p. 141.
134. *Buck v. Bell*, 274 U.S. 200 (1927), p. 207.
135. For more information on Francis Galton, see Haller, *Eugenics*, pp. 8–20, and Daniel J. Kevles, *In the Name of Eugenics: Genetics and the Uses of Human Heredity* (Cambridge: Harvard University Press, 1995), pp. 3–19.
136. Kevles, *Eugenics*, p. xiii.
137. Edwin Black, *War Against the Weak: Eugenics and America's Campaign to Create a Master Race* (New York: Four Walls Eight Eindows, 2003), p. 12.
138. "Understanding Evolution," http://evolution.berkeley.edu/evosite/misconceps/IIIBmight.shtml (accessed August 26, 2006).
139. David Micklos, "None Without Hope: *Buck vs. Bell* at 75," http://www.dnalc.org/resources/buckvbell.html (accessed August 15, 2003). Micklos is director of the DNA Learning Center at the Cold Spring Harbor Laboratory, Long Island,

New York.

140. *Descent* (1871), vol. I, p. 168.

141. Ibid., pp. 168–169. p. 113.

142. Ibid., pp. 172–173.

143. Ibid., p. 174.

144. Ibid., pp. 177–178.

145. *Descent* (1896), vol I., p. 143.

146. Wallace, "Human Selection," in *Alfred Russel Wallace, An Anthology of His Shorter Writings*, ed. by Charles H. Smith (New York: Oxford University Press, 1991), p. 51. This essay by Wallace was published in 1890, after Darwin's death.

147. *Descent* (1871), vol. II, p. 403. Also see John Bowlby, *Darwin: A New Life* (New York: W.W. Norton, 1990), pp. 415–16.

148. "Had he not been subjected to natural selection, assuredly he would never have attained to the rank of manhood." *Descent* (1871), vol. I, p. 180.

149. Ibid., vol. II, p. 403.

150. Frank Lowden, "Social Work in Government," *Proceedings of the National Conference of Social Work*, May 16–23, 1923 (Chicago: University of Chicago Press), pp. 150–151.

151. Edwin Conklin, "Value of Negative Eugenics," *Journal of Heredity*, vol VI, no. 12 (December 1915), pp. 539–540.

152. Paul Popenoe and Roswell Johnson, "Eugenics and Euthenics" in Horatio Hackett Newman, *Evolution, Genetics and Eugenics*, third edition (Chicago: University of Chicago Press, 1932), pp. 517–518.

153. H. E. Jones, "Heredity as a Factor in the Improvement of Social Conditions," *American Breeders Magazine*, vol II, no. 4. (Fourth Quarter, 1911), p. 253.

154. Davenport, "Scientific Cooperation with Nature: Eugenics," typescript, Charles Davenort Papers, Manuscript Collection B D27, American Philosophical Society.

155. Edward M. East, *Heredity and Human Affairs* (New York: Charles Scribner's Sons, 1927), pp. 307, 306.

156. Edward East was a professor at Harvard. Edwin Conklin was a professor at Princeton. Henry Fairfield Osborn was a professor at Columbia. And David Starr Jordan was president of Stanford University.

157. For information about colleges and universities where eugenics and eugenics related courses were taught, see folders on "Eugenics Questionnaire," in the Eugenics Record Office papers, Manuscript Collection 77, American Philosophical Society. For treatments of eugenics in secondary school biology textbooks of the period, see Clifton Hodge and Jean Dawson, *Civic Biology* (Boston: Ginn and Company, 1918), 344–345; William Smallwood, Ida Reveley, and Guy Bailey, *New Biology* (Boston: Allyn and Bacon, 1924), 660–662; George William Hunter, *New Civic Biology* (New York: American Book Company, 1926), 251, 398–403. For a general summary of the coverage of eugenics in secondary school biology texts, see Gerald Skoog, "Topic of Evolution in Secondary School Biology Textbooks: 1900–1977," *Science Education* 63, no. 5 (1979): 628. Some textbooks during this period did not promote eugenics, or only did so indirectly. Alfred Kinsey's *An*

Introduction to Biology (Philadelphia: J. B. Lippincott, 1926) had no discussion of eugenics per se, but it did recommend books advocating eugenics in a list of reference books supplied on pp. 539–540. By the 1940s, discussions of eugenics were significantly diluted and focused more on environment than heredity. For an example of a later treatment of the topic, see Edwin Sanders, *Practical Biology* (New York: D. Van Nostrand. Co, 1947), 431–437. For examples of college biology textbooks that covered eugenics, see George William Hunter, Herbert Eugene Walter, and George William Hunter, III, *Biology: The Study of Living Things* (New York: American Book Company, 1937), 638–642; Arthur Haupt, *Fundamentals of Biology* (New York: McGraw-Hill Book Company, 1928), 216–221; Leslie Kenoyer and Henry Goddard, *General Biology* (New York, Harper and Brothers, 1937), 533–539; William Martin Smallwood, *A Text-Book of Biology for Students in Medical, Technical and General Courses* (Philadelphia: Lea and Febiger, 1913), 257–259. Not every college textbook during this period jumped on the eugenics bandwagon. Waldo Shumway's *Textbook of General Biology* (New York: John Wiley and Sons, 1931) told students that "it would appear judicious to walk slowly in enacting [eugenics] legislation until our knowledge of biological laws is more precise," p. 317.

158. East, *Heredity and Human Affairs*, p. 237.
159. Edwin Grant Conklin, *Heredity and Environment in the Development of Men*, fifth edition revised (Princeton: Princeton University Press, 1923), p. 295; also, more generally, pp. 292–315. For similar thoughts in other editions, see Conklin, *Heredity and Environment in the Development of Men* (Princeton: Princeton University Press, 1915), pp. 404–440; Conklin, *Heredity and Environment in the Development of Men*, second edition (Princeton: Princeton University Press, 1916), pp. 416–453; Conklin, *Heredity and Environment in the Development of Men*, third edition (Princeton: Princeton University Press, 1919), pp. 275–297.
160. Conklin was also supportive of the idea of positive eugenics. "No eugenical reform can fail to take account of the fact that the decreasing birth rate among intelligent people is a constant menace to the race. We need not 'fewer and better children' but more children of the better sort and fewer of the worse variety." [Conklin, *Heredity and Environment* (1923), p. 314.] He went on to criticize birth control reformers for believing that "the race is to be regenerated through sterilization or birth control. But unfortunately this reform begins among those who because of good hereditary traits should not be infertile." [Ibid.]
161. Arnhart, *Darwinian Conservatism*, p. 122.
162. Frederick Engels, *Ludwig Feuerbach and the End of Classical German Philosophy* (originally published in 1886; Progress Publishers edition, 1946), part IV; available online at http://www.marxists.org/archive/marx/works/1886/ludwig-feuerbach/index.htm (accessed June 23, 2005).
163. Karl Marx to Ferdinand Lassalle, Jan. 16, 1861, *Karl Marx, Frederick Engels: Collected Works* (New York: International Publishers, 1975–2005), vol. 41, p. 245.
164. Engels went on to add: "Never before has so grandiose an attempt been made to demonstrate historical evolution in Nature, and certainly never to such good effect. One does, of course, have to put up with the crude English method." [Frederick Engels to Karl Marx, December 11 or 12, 1859, *Collected Works*, vol.

40, p. 550.] The views of Marx and Engels on Darwin's death-blow to teleology were echoed by other German socialists. "German socialists... were elated with Darwin's elimination of teleology from nature, which they regularly summoned in defense of their materialist world view." [Richard Weikart, *Socialist Darwinism: Evolution in German Social Thought from Marx to Bernstein* (San Francisco: International Scholars Publications, 1999), p. 221.] The same view was adopted by later Soviet thinkers. "In Soviet opinion this theory of Darwin's deals a devastating blow against the teleological point of view." [Gustav A. Wetter, *Dialectical Materialism: A Historical and Systematic Survey of Philosophy in the Soviet Union*, trans. by Peter Heath (New York: Frederick A. Praeger, Publishers, 1958), pp. 379–380.]

165. Karl Marx to Frederick Engels, December 19, 1860, *Collected Works*, vol. 41, p. 232.
166. Karl Marx to Ferdinand Lassalle, January 16, 1861, *Collected Works*, vol. 41, p. 245.
167. Weikart, *Socialist Darwinism*, p. 35. Despite his praise of Darwin, Marx was ambivalent about the application of Darwinism to human society, and he could be very negative about Darwin's theory in private. Ibid., pp. 15–51. For an interesting summary of the contacts between Darwin and Marx, see Ralph Colp, Jr., "The Contacts between Karl Marx and Charles Darwin," *Journal of the History of Ideas*, vol. 35, No. 2 (April–June 1974), pp. 329–338. For many years it was believed that Marx offered to dedicate an edition of *Das Kapital* to Darwin, but this idea was based on a misidentification of the recipient of one of Darwin's letters. See Margaret A. Fay, "Did Marx Offer to Dedicate *Capital* to Darwin?: A Reassessment of the Evidence," *Journal of the History of Ideas*, vol. 39, no. 1 (January–March, 1978), pp. 133–146.
168. Bannister, *Social Darwinism*, pp. 133–135; Weikart, *Socialist Darwinism*, pp. 83–101.
169. Frederick Engels to Pyotr Lavrov, Nov. 12–17, 1875, *Collected Works*, vol. 45, p. 106.
170. See Weikart, *Socialist Darwinism*, pp. 4–5, 15–82, 221–222.
171. Gertrude Himmelfarb, *Darwin and the Darwinian Revolution* (Chicago: Elephant Paperbacks, 1996), p. 422.
172. As Marx put it, "philosophers have hitherto only interpreted the world... the point is to change it." Karl Marx, "Theses On Feuerbach" (1845), trans. by Cyril Smith, http://www.marxists.org/archive/marx/works/1845/theses/ (accessed June 24, 2005).
173. For a discussion of the impact of Germanic political science on America during the Progressive era, see Dennis John Mahoney, "A New Political Science for a World Made Wholly New: The Doctrine of Progress and the Emergence of American Political Science," (Ph.D. Dissertation, Claremont Graduate School, 1984), especially pp. 25–45. Mahoney discusses the evolutionary character of Hegelian political science on pp. 33–35.
174. See Mahoney, "A New Political Science," pp. 103–142.
175. Woodrow Wilson, *Constitutional Government in The Papers of Woodrow Wilson*,

Arthur S. Link, editor (Princeton: Princeton University Press, 1974), volume 18, p. 105.

176. Woodrow Wilson, *The New Freedom*, pp. 41–42.

VI. Is Darwinianism Compatible with Religion?

177. The first quote comes from Richard Dawkins, "Is Science a Religion?" http://www.thehumanist.org/humanist/articles/dawkins.html (accessed July 16, 2005); the second quote is from Richard Dawkins, *The Blind Watchmaker: Why the Evidence of Evolution Reveals a Universe Without Design* (New York: W.W. Norton and Co., 1996), p. 6.
178. Arnhart, *Darwinian Conservatism*, p. 85.
179. James Q. Wilson, "Debating Intelligent Design," [letter to the editor], *The Claremont Review of Books*, Summer 2006, p. 7.
180. Krauthammer, "Phony Theory."
181. See "Misconception: Evolution and religion are incompatible," http://evolution.berkeley.edu/evosite/misconceps/IVAandreligion.shtml;Statements from Religious Organizations (Oakland, California: National Center for Science Education), http://www.ncseweb.org/resources/articles/5025_statements_from_religious_orga_12_19_2002.asp#home (accessed July 16, 2005); Eugenie Scott, "Dealing with Antievolutionism," http://www.ucmp.berkeley.edu/fosrec/Scott2.html (accessed August 25, 2006).
182. In her tips for activists who want to support evolution before their school board, Eugenie Scott advises: "Call on the clergy. Pro-evolution clergy are essential to refuting the idea that evolution is incompatible with faith.... If no member of the clergy is available to testify, be sure to have someone do so—the religious issue must be addressed in order to resolve the controversy successfully." Eugenie C. Scott, "12 Tips for Testifying at School Board Meetings," http://www.ncseweb.org/resources/articles/7956_12_tips_for_testifying_at_scho_3_19_2001.asp (accessed July 16, 2005), emphasis in original.
183. *Congregational Study Guide for Evolution* (Oakland, California: National Center for Science Education, 2001), http://www.ncseweb.org/article.asp?category=11 (accessed July 16, 2005).
184. Phina Borgeson, "Introduction to the Congregational Study Guide for Evolution," http://www.ncseweb.org/resources/articles/8888_csg-int.pdf (accessed July 16, 2005).
185. For information about these initiatives, see http://www.butler.edu/clergyproject/clergy_project.htm (accessed August 25, 2006).
186. Richard Dawkins, "Is Science a Religion?" http://www.thehumanist.org/humanist/articles/dawkins.html (accessed July 16, 2005).
187. New Orleans Secular Humanist Association home page, http://nosha.secularhumanism.net/index.html. Forrest is listed as a member of the board of directors on the "Who's Who" page of the website, http://nosha.secularhumanism.net/whoswho.html (accessed July 6, 2002).
188. Victor Stenger, Has Science Found God?, preface, http://www.colorado.edu/phi-

losophy/vstenger/Found/00Preface.pdf, p. 16 (accessed August 26, 2006).

189. Steven D. Schafersman, "The History and Philosophy of Humanism and its Role in Unitarian Universalism," speech originally delivered in Sept. 1995 and revised in December 1998, http://www.freeinquiry.com/humanism-uu.html (accessed July 25, 2005).
190. Steven Schafersman, "The Challenge of the Fossil Record," http://www.freeinquiry.com/challenge.html (accessed July 25, 2005).
191. Steven Weinberg, quoted in "Free People from Superstition," http://ffrf.org/fttoday/2000/april2000/weinberg.html (accessed July 25, 2005).
192. "Humanism and Its Aspirations: Humanist Manifesto III," (Washington, D.C.: American Humanist Association), http://www.americanhumanist.org/3/HumandItsAspirations.htm (accessed July 16, 2005).
193. See Larry Witham, *Where Darwin Meets the Bible: Creationists and Evolutionists in America* (New York: Oxford University Press, 2002), pp. 271–273.
194. The proportion of Americans who believe in the existence of God is 80–94%, depending on how the question is asked. In a 2004 Gallup Poll, 89.9% of respondents said they believed in God. When offered a choice between God or "a universal spirit or higher power," 80.9% said they believed in God, while another 12.6% said they believed in a universal spirit or higher power, for a combined rate of 93.5%. This is virtually the same result as a Gallup Poll in 1988, which found that 94.6% of respondents said they believed "in God or a universal spirit." [Gallup Poll, 5/02/2004–5/04/2004, 12/21/1988–12/22/1988, http://brain.gallup.com/ (accessed July 16, 2005] The proportion of Americans who believe in life after death is 68–77%, depending on how the question is asked. In a 1988 Gallup Poll, 68.1% of respondents said they believed in "life after death," but 76.8% of respondents said that "there is a Heaven where people who had led good lives are eternally rewarded." [Gallup Poll, 12/21/1988–12/22/1988, http://brain.gallup.com/ (accessed July 16, 2005)]
195. "Darwin Day 2002 Events Calendar," http://www.darwinday.org/events/calendar.html (accessed June 6, 2002).
196. See Bruce Chapman, "Vatican Astronomer Replaced," http://www.evolutionnews.org/2006/08/vatican_astronomer_replaced.html (accessed August 24, 2006).
197. George V. Coyne, S. J., "The Dance of the Fertile Universe," p. 7, available at http://www.aei.org/docLib/20051027_HandoutCoyne.pdf (accessed August 25, 2006).
198. For an account of the lecture, see "From Calvinism to Freethought: The Road Less Traveled," http://www.freethoughtassociation.org/minutes/2006/May24-2006.htm (accessed July 25, 2006).
199. Arnhart, *Darwinian Conservatism*, p. 91.
200. Ibid., p. 89.
201. Peter J. Bowler, *Darwinism* (New York: Twayne Publishers, 1993), p. 6.
202. Ronald L. Numbers, *Darwinism Comes to America* (Cambridge, MA: Harvard University Press, 1998), p. 27.
203. See discussion in Edward Larson, *Summer for the Gods: The Scopes Trial and America's Continuing Debate Over Science and Religion* (New York: Basic Books,

1997), pp. 19–26.

204. Francis S. Collins, *The Language of God: A Scientist Presents Evidence for Belief* (New York: Free Press, 2006), p. 205.
205. Psalm 19:1 (*NKJV*).
206. Romans 1:20 (*NKJV*).
207. Kenneth R. Miller, *Finding Darwin's God: A Scientist's Search for Common Ground Between God and Evolution* (New York: HarperCollins, 1999), p. 244.
208. Ibid., p. 272. Despite Miller's clearly stated view that evolution is "undirected," Miller also claims in apparent contradiction that "the final result of the process may nonetheless be seen as part of God's will" [p. 236]. What Miller seems to mean by this is that once God set up the undirected and unpredictable process of evolution, he could know that "given evolution's ability to adapt, to innovate, to test, and to experiment, sooner or later it would have given the Creator exactly what He was looking for—a creature who, like us, could know Him and love Him."[pp. 238–239] By saying that evolution "would have given the Creator exactly what He was looking for" Miller is engaging in word games. In fact, according Miller's view human beings do not represent an "exact" intention of God, at least in any way that most people would commonly understand that term. In Miller's view, while God may have wished for some sort of rational creature to develop in the universe, he assigned the job to an undirected process that could have produced any number of different results other than human beings. Thus, it was mere "happenstance" that human beings developed. Miller's view is a radical departure from traditional Christian (or Jewish or Muslim) teaching that human beings are created as the result of God's specific plan. The point here is not to argue whether Miller's view or the view of traditional theology is correct, but to point out that Miller's theological defense of unguided evolution is open to significant challenge from traditional Judeo-Christian theology.

VII. Has Darwinism Refuted Intelligent Design?

209. *Kitzmiller et. al. v. Dover Area School Board* (M.D. Pa. December 20, 2005), slip opinion, p. 89.
210. Answer to "What is the theory of intelligent design?" http://www.discovery.org/csc/topQuestions.php - questionsAboutIntelligentDesign (accessed August 10, 2006).
211. Arnhart, *Darwinian Conservatism*, p. 101.
212. For an explanation of the Institute's science education policy, see "Questions about Science Education Policy," http://www.discovery.org/csc/topQuestions.php - questionsAboutScienceEducationPolicy (accessed August 10, 2006). For a discussion of why the Institute opposes mandating intelligent design, see "Discovery Institute Opposes Proposed PA Bill on Intelligent Design" (June 22, 2005), http://www.discovery.org/scripts/viewDB/index.php?command=view&id=2688 (accessed August 10, 2006).
213. For information about the Institute's opposition to the Dover policy, see "Setting the Record Straight about Discovery Institute's Role in the Dover School District Case" (Nov. 10, 2005), http://www.discovery.org/scripts/viewDB/index.

php?command=view&id=3003 (accessed August 10, 2006). Arnhart in his paper prepared for the American Political Science Association erroneously claims that "[i]nitially, the Discovery Institute supported them [the Dover board] and arranged to provide expert witnesses for them. But then shortly before the trial began, the Discovery Institute announced that it opposed the policy of the school board." Larry Arnhart, "Darwinian Conservatism," paper prepared for presentation at the American Political Science Association annual meeting, August 31, 2006, Philadelphia, Pennsylvania, pp. 48–49. In fact, Discovery Institute publicly opposed the Dover policy soon after it was adopted, and this opposition was mentioned in an Associated Press report in early November 2004. [See "AP Cites Discovery Institute's Opposition to Dover School Board Policy" (Nov. 12, 2004), http://www.discovery.org/scripts/viewDB/index.php?command=view&id=2848 (accessed August 10, 2006).] Then in December 2004 the Institute publicly urged the policy's repeal (it had privately urged its repeal between October and December 2004). [See "Discovery Calls Dover Evolution Policy Misguided, Calls For its Withdrawal" (Dec. 14, 2005), http://www.discovery.org/scripts/viewDB/index.php?command=view&id=2341 (acccessed August 10, 2006).] After the school board refused to repeal its policy, and it was sued, Discovery Institute did agree to assist with expert witnesses because it saw that the ACLU was trying to place intelligent design itself on trial, not just the school district policy. Two Discovery Fellows ultimately testified at the trial, while several others did not. For further information, see "Setting the Record Straight."

214. Arnhart, *Darwinian Conservatism*, p. 102.
215. Arnhart, "Darwinian Conservatism" (2006, paper; for full reference, see note 212), p. 47.
216. Ibid.
217. See http://www.dissentfromdarwin.org/ (accessed August 9, 2006); "80 Years After Scopes Trial New Scientific Evidence Convinces Over 400 Scientists That Darwinian Evolution is Deficient," July 18, 2005, http://www.discovery.org/scripts/viewDB/index.php?command=view&id=2732 (accessed July 25, 2005). This statement was originally issued in 2001; see "A Scientific Dissent from Darwinism," http://www.reviewevolution.com/press/DarwinAd.pdf (accessed July 25, 2005).
218. Quoted in "100 Scientists, National Poll Challenge Darwinism" (Seattle: Discovery Institute, September 24, 2001), http://www.reviewevolution.com/press/pressRelease_100Scientists.php (accessed July 25, 2005).
219. Quoted in "80 Years After Scopes Trial."
220. ColdWater Media Interview with Scott Minnich, 2001.
221. Quoted in "40 Texas scientists join growing national list of scientists skeptical of Darwin," September 5, 2003, http://www.discovery.org/scripts/viewDB/index.php?command=view&id=1555 (accessed July 26, 2005).
222. Andrew Welsh-Huggins (Associated Press), "State school board approves evolution lesson plan," March 10, 2004, http://www.ohio.com (accessed March 10, 2004); Glen Needham, "Science Standards will Spur Critical Thinking," *The Columbus Dispatch*, March 1, 2004, A-6.
223. Needham, "Science Standards." The model lesson was repealed in 2006 after

changes on the Board and a lobbying campaign by evolutionists in Ohio.

224. Jodi Wilgoren, "Kansas Begins Hearings on Diluting Teaching of Evolution," May 5, 2005, *The New York Times*.

225. For a complete list of witnesses at the Kansas hearings, see http://www.kansas-science2005.com/WitnessesScienceHearings.pdf (accessed August 9, 2006).

226. See "South Carolina Set to Join Four Other States Calling for Critical Analysis of Evolution" (June 8, 2006), http://www.discovery.org/scripts/viewDB/index.php?command=view&id=3523 (accessed August 9, 2006); "South Carolina Praised for Requiring Students to Critically Analyze Evolutionary Theory" (June 12, 2006), http://www.discovery.org/scripts/viewDB/index.php?command=view&id=3527 (accessed August 9, 2006).

227. "Brief of Amici Curiae Biologists And Other Scientists In support of Appellants," filed in the case of Cobb County School District v. Selman, p. 6, available online at http://www.discovery.org/scripts/viewDB/filesDB-download.php?command=download&id=619 (accessed August 9, 2006).

228. For examples of materials submitted to educational policymakers citing peer-reviewed research on these issues, see "Bibliography of Supplementary Resources"; "The Scientific Controversy over Whether Microevolution Can Account for Macroevolution," (Seattle: Discovery Institute), http://www.discovery.org/scripts/viewDB/filesDB-download.php?id=118 (accessed July 25, 2005); "The Scientific Controversy over the Cambrian Explosion," (Seattle: Discovery Institute), http://www.discovery.org/scripts/viewDB/filesDB-download.php?id=119 (accessed July 25, 2005); "An Analysis of the Treatment of Homology in Biology Textbooks," (Seattle: Discovery Institute), http://www.discovery.org/articleFiles/PDFs/homologyrpt.pdf (accessed July 25, 2005); "A Preliminary Analysis of the Treatment of Evolution in Biology Textbooks," (Seattle: Discovery Institute), http://www.discovery.org/articleFiles/PDFs/TexasPrelim.pdf (accessed July 25, 2005).

229. "Six Myths about the Evolution Debate" (Discovery Institute, August 2005), http://www.discovery.org/scripts/viewDB/filesDB-download.php?command=download&id=476 (accessed August 9, 2006).

230. Schafersman initially claimed that current biology textbooks were already completely "factually accurate and free of errors concerning evolution… nor do they omit scientific information critical of evolution, because there isn't any such information." [Written testimony of Dr. Steven Schafersman submitted to the Texas State Board of Education for the public hearing on textbooks on July 9, 2003.]

231. Revised version of the written testimony of Dr. Steven Schafersman submitted to the Texas State Board of Education for the public hearing on textbooks on July 9, 2003, available online at http://www.texscience.org/files/schafersman-july9-testimony.htm (accessed July 25, 2005). This version differs in certain respects from the written testimony Dr. Schafersman actually submitted to the Board of Education.

232. Written testimony of Dr. Steven Schafersman submitted to the Texas State Board of Education for the public hearing on textbooks on September 10, 2003, available online at http://www.texscience.org/files/icons-revealed/index.htm (accessed July 25, 2005).

233. Arnhart, *Darwinian Conservatism*, p. 97.

234. Ibid.

235. Ibid.

236. Ibid.

237. Krauthammer, "Phony Theory."

238. See William Dembski, *The Design Inference: Eliminating Chance through Small Probabilities* (Cambridge University Press, 1998); and William Dembski, *No Free Lunch: Why Specified Complexity Cannot Be Purchased without Intelligence* (Rowman and Littlefield, 2002).

239. Arnhart, *Darwinian Conservatism*, p. 97.

240. Charles B. Thaxton, "A Word to the Teacher," in Percival Davis and Dean H. Kenyon, *Of Pandas and People: The Central Question of Biological Origins* (Dallas: Haughton Publishing Co., 1989), pp. 156–157.

241. Michael Behe, "Molecular Machines: Experimental Suport for the Design Inference" (March 1, 1998), http://www.discovery.org/scripts/viewDB/index.php?command=view&id=54(accessed August 9, 2006).

242. See Stephen Meyer, "DNA and the Origin of Life: Information, Specification, and Explanation," in *Darwinism, Design, and Public Education*, pp. 266–270;

243. See William Dembski, "Is Intelligent Design Testable? A Response to Eugenie Scott" (January 24, 2001), http://www.discovery.org/scripts/viewDB/index.php?command=view&id=584 (accessed August 9, 2006); Jay Richards and Jonathan Witt, "Intelligent Design is Empirically Testable and Makes Predictions" (January 5, 2006), http://www.evolutionnews.org/2006/01/intelligent_design_is_empirica.html (accessed August 9, 2006).

244. Wilson, "Debating Intelligent Design."

245. Arnhart, *Darwinian Conservatism*, p. 96.

246. Michael Behe, *Darwin's Black Box: The Biochemical Challenge to Evolution* (New York: The Free Press, 1996), p. 40.

247. Arnhart, "Darwinian Conservatism" (2006, paper; for full reference, see note 212), p. 41.

248. Scott Minnich and Stephen C. Meyer, "Genetic Analysis of Coordinate Flagellar and Type III Regulatory Circuits," Proceedings of the Second International Conference on Design & Nature, edited by M. W. Collins and C.A. Brebbia (Wessex Institute of Technology Press, 2004).

249. Ibid.

250. James A. Shapiro, "In the details . . . what?" *National Review* (Sept. 16, 1996), pp. 62–65.

251. Franklin M. Harold, *The Way of the Cell: Molecules, Organisms and the Order of Life* (Oxford University Press, 2001), p. 205.

252. Heinz-Albert Becker & Wolf-Ekkehard Lönnig, "Transposons: Eukaryotic," *Encyclopedia of Life Sciences* (2001), p. 8.

253. Evelyn Fox Keller, "Developmental Robustness," *Ann. N.Y. Acad. Sci.*, 981 (2002), pp. 189, 190, emphasis added.

254. Stephen C. Meyer, "The Origin of Biological Information in the Higher Taxonomic Categories," *Proceedings of the Biological Society of Washington*, 117 (2)

(2004): 213–239.

255. Michael J. Behe and David W. Snoke, "Simulating Evolution by Gene Duplication of Protein Features That Require Multiple Amino Acid Residues," *Protein Science*, 13 (2004): 2651–2664.

256. D. Axe, "Estimating the prevalence of protein sequences adopting functional enzyme folds, *J. Mol. Biol.* 341 (2004): 1295–1315; D. Axe, "Extreme functional sensitivity to conservative amino acid changes on enzyme exteriors," *Journal of Molecular Biology*, 301 (2000): 585–596. Meyer summarizes the importance of Axe's findings thus: "Since natural selection can do nothing to help generate new functional sequences, but rather can only preserve such sequences once they have arisen, chance alone—random variation—must do the work of information generation—that is, of finding the exceedingly rare functional sequences within the set of combinatorial possibilities. Yet the probability of randomly assembling (or 'finding,' in the previous sense) a functional sequence is extremely small... [According to Axe's experiments,] the probability of finding a functional protein among the possible amino acid sequences corresponding to a 150-residue protein is... 1 in 10^{77}." Meyer, "The Origin of Biological Information."

257. Quoted in Darry Madden, "UMass scientist to lead debate on evolutionary theory," *Brattleboro Reformer*, Feb. 2, 2006, http://www.reformer.com/Stories/0,1413,102~8862~3224997,00.html (accessed February 2, 2006).

258. Arnhart, *Darwinian Conservatism*, p. 99.

259. Quoted in Adam Wolfson, "Survival of the Evolution Debate: Why Darwin is still a lightning rod," *The Weekly Standard*, January 16, 2006, vol. 11, issue 17.

260. According to Will, "[t]oday's proponents of 'intelligent design' theory are doing nothing novel when they say the complexity of nature is more plausibly explained by postulating a designing mind—a.k.a. God—than by natural adaptation and selection... intelligent design... is not a scientific but a creedal tenet—a matter of faith, unsuited to a public school's science curriculum." [George F. Will, "A Debate that Does Not End," *Newsweek*, July 4, 2005.] According to Krauthammer, "[i]ntelligent design may be interesting as theology, but as science it is a fraud. It is a self-enclosed, tautological 'theory' whose only holding is that when there are gaps in some area of scientific knowledge—in this case, evolution—they are to be filled by God." [Krauthammer, "Phony Theory."] According to Wilson, "I.D. supporters repeatedly argue that [scientific] facts prove God's existence and thus evolution must be wrong." [Wilson, "Debating Intelligent Design."] John Derbyshire, meanwhile, praises the following comment from a reader and cites it as "an appropriate valediction" to his discussion of intelligent design: "Now to the dishonesty of IDers. Intelligent Design supposes that supernatural forces have crafted the world as we see it. Supernatural forces are simply not within the scope of science. Science necessarily only concerns itself with natural phenomena and natural causes. Supernatural causes are not testable, quantifiable, or qualifiable. They are simply not the scope of science. ID is unscience." [John Derbyshire, "The Last Word," http://www.nationalreview.com/thecorner/05_02_06_corner-archive.asp - 055887 (accessed September 12, 2006)]. Inconsistently, less than a month earlier Derbyshire himself had conceded that Michael Behe's argument for intelligent design "does not necessarily imply **supernatural** intervention."

[John Derbyshire, "Re: Atheists for Intelligent Design," http://www.nationalreview.com/thecorner/05_01_09_corner-archive.asp - 049839(accessed September 12, 2006).]

261. *Kitzmiller v. Dover*, slip opinion, p. 30.
262. Arnhart, *Darwinian Conservatism*, p. 99.
263. Charles B. Thaxton, "In Pursuit of Intelligent Causes: Some Historical Background," a lecture originally delivered in 1988, available at http://www.origins.org/articles/thaxton_pursuitofintelligent.html (accessed August 23, 2006).
264. For documentation of this point, see "Appendix A to Amicus Brief filed by Discovery Institute in Tammy J. Kitzmiller et al. v. Dover Area School District and Dover Area School District Board of Directors, Civil Action No. 4:04-cv-2688: Documentation showing that the scientific theory of intelligent design makes no claims about the identity or nature of the intelligent cause responsible for life," available at http://www.discovery.org/scripts/viewDB/filesDB-download.php?command=download&id=647 (accessed August 23, 2006).
265. Percival Davis and Dean H. Kenyon, *Of Pandas and People: The Central Question of Biological Origins* (Dallas: Haughton Publishing Co., 1989), p. 7; Percival Davis and Dean H. Kenyon, *Of Pandas and People: The Central Question of Biological Origins*, second edition (Dallas: Haughton Publishing Co., 1993), p. 7.
266. Ibid., pp. 126–127 of both editions. Also see p. 161 of 1993 edition.
267. Michael Behe, "The Modern Intelligent Design Hypothesis," *Philosophia Christi*, series 2, vol. 3, no. 1 (2001), p. 165.
268. Arnhart, *Darwinian Conservatism*, p. 99.
269. William Dembski, *Intelligent Design: The Bridge Between Science and Theology* (Downers Grove, IL: InterVarsity Press, 1999), pp. 247–249.
270. For a good discussion of the debate over the definition of science and its applicability to intelligent design, see Stephen C. Meyer, "The Scientific Status of Intelligent Design: The Methodological Equivalence of Naturalistic and Non-Naturalistic Origins Theories," in Michael J. Behe, William A. Dembski, and Stephen C. Meyer, *Science and Evidence for Design in the Universe* (San Francisco: Ignatius Press, 2000), pp. 151–211. Also see David DeWolf, John West, Casey Luskin, and Jonathan Witt, *Traipsing Into Evolution: Intelligent Design and the Kitzmiller vs. Dover Decision* (Seattle: Discovery Institute Press, 2006), pp. 25–28.
271. Meyer, "The Scientific Status of Intelligent Design."
272. Ernst Mayr, "Darwin's Influence on Modern Thought," *Scientific American*, July 2000, p. 80.

Conclusion

273. G. Sermonti, "Darwin is a Prime Number," *Rivista di Biologia*, 95 [2002], p. 10.
274. Krauthammer, "Phony Theory."
275. Will, "Grand Old Spenders."
276. Krauthammer, "Phony Theory."
277. "Rationale of the State Board for Adopting these Science Curriculum Standards,"

Kansas Science Education Standards, adopted November 2005 by the Kansas State Board of Education, emphasis added.

278. Will and Krauthammer also falsely claim that the Kansas Board redefined science in order to include the supernatural, when in fact the Board was reinstating the standard definition of science used in virtually all other states. See Jonathan Wells, "Analysis of Kansas Definition of Science Compared to All Other State Science Definitions," http://www.discovery.org/scripts/viewDB/index.php?command=view&id=2573 (accessed September 6, 2006).
279. Will, "A Debate that Does Not End."
280. See "Intelligent Design is Falsifiable," http://www.discovery.org/scripts/viewDB/index.php?command=view&id=2812 (accessed September 6, 2006); also see Dembski, "Is Intelligent Design Testable?"; Richards and Witt, "Intelligent Design is Empirically Testable and Makes Predictions."
281. Stephen M. Barr, "The Design of Evolution," *First Things*, October 2005, http://www.firstthings.com/ftissues/ft0510/opinion/barr.html (accessed August 26, 2006).
282. Darwin, *Variation of Animals and Plants under Domestication*, pp. 428–429
283. Carson Holloway, *The Right Darwin? Evolution, Religion, and the Future of Democracy* (Dallas: Spence Publishing, 2006), p. 185.
284. Arnhart, *Darwinian Conservatism*, p. 135.
285. Francis Darwin, editor, *The Autobiography of Charles Darwin and Selected Letters* (New York: Dover Publications, 1958), p. 65.
286. Holloway, *The Right Darwin*, p. 185.
287. Jonah Goldberg, "Evolution, Kansas, NPR, Intelligent Design," http://www.nationalreview.com/thecorner/05_02_06_corner-archive.asp - 055600 (accessed September 12, 2006).
288. Harry V. Jaffa, "Debating Intelligent Design," *The Claremont Review of Books*, Summer 2006, p. 9.
289. See, for example, Chris Mooney, *The Republican War on Science* (New York: Basic Books, 2005).
290. http://www.fenton.com/pages/3_ourwork/1_clients/clients.htm (accessed September 9, 2006).
291. See http://www.defconamerica.org/ (accessed September 9, 2006); "Introducing the Campaign to Defend the Constitution: Because the Religious Right is Wrong," *Fenton Communique*, Spring 2006, p. 3, available at http://www.fenton.com/pages/5_resources/pdf/Spring2006_newsletter.pdf. For more information about Fenton Communications, see Chuck Colson, "Bankrolling Hostility," Breakpoint commentary, September 8, 2006, http://www.breakpoint.org/listingarticle.asp?ID=2432 (accessed September 9, 2006).
292. "Eugenics in Politics," *The New York Times*, October 9, 1921, p. 93.
293. John Derbyshire, "The Darwinian Mandarinate," http://www.nationalreview.com/thecorner/05_02_06_corner-archive.asp - 055833 (accessed September 11, 2006); John Derbyshire, "Scientists Libertarian," http://www.nationalreview.com/thecorner/05_02_06_corner-archive.asp - 055846 (accessed September 11,

2006).

294. C. S. Lewis, "Is Progress Possible? Willing Slaves of the Welfare State," in *God in the Dock: Essays on Theology and Ethics* (Grand Rapids: Eerdmans, 1970), p. 315.

295. Wilson, "Debating Intelligent Design."

296. See "Intelligent Design and Academic Freedom," on *All Things Considered*, National Public Radio, November 10, 2005; Geoff Brumfiel, "Intelligent design: Who has designs on your students' minds?" *Nature* (April 28, 2005), volume 434, pp. 1062–1065, available online at http://www.nature.com/nature/journal/v434/n7037/full/4341062a.html (accessed July 26, 2005).

297. Letter to Richard Sternberg from the U.S. Office of Special Counsel, August 5, 2005, available at http://www.rsternberg.net/OSC_ltr.htm (accessed September 9, 2006). Also see David Klinghoffer, "The Branding of a Heretic," *The Wall Street Journal*, January 28, 2005, http://www.opinionjournal.com/taste/?id=110006220 (accessed July 26, 2005). For more information about the controversy surrounding the publication of the journal article supportive of intelligent design, see "Info on Meyer Article Published in Smithsonian Journal," http://www.discovery.org/scripts/viewDB/index.php?command=view&id=2399 (accessed July 26, 2005), and http://www.rsternberg.net/ (accessed July 26, 2005).

298. Quoted in Michael Powell, "Editor Explains Reasons for 'Intelligent Design' Article," *The Washington Post*, August 19, 2005, A19.

299. Testimony of Nancy Bryson before the Texas State Board of Education, September, 2003, *Transcript of the Public Hearing before the Texas State Board of Education*, September 10, 2003, Austin, Texas (Austin, Texas: Chapman Court Reporting Service, 2003), pp. 504–505.

300. For information about the Bryan Leonard case, see Catherine Candinsky, "Evolution debate re-emerges: Doctoral student's work was possibly unethical, OSU professors argue," *The Columbus Dispatch*, June 9, 2005, http://www.dispatch.com/news-story.php?story=dispatch/2005/06/09/20050609-C1-01.html&chck=t>http://www.dispatch.com/news-story.php?story=dispatch/2005/06/09/20050609-C1-01.html&chck=t (accessed June 9, 2005); "Attack on OSU Graduate Student Endangers Academic Freedom," http://www.discovery.org/scripts/viewDB/index.php?command=view&id=2661 (accessed July 26, 2005); "Professors Defend Ohio Grad Student Under Attack by Darwinists," http://www.discovery.org/scripts/viewDB/index.php?command=view&id=2715 (accessed July 26, 2005).

301. For examples, see Brian Leiter, "Biology Textbooks under Attack," August 11, 2003, *The Leiter Reports*, http://webapp.utexas.edu/blogs/archives/bleiter/000146.html#000146 (accessed July 25, 2005); "Statement of the Reverend Mark Belletini," Ohio Citizens for Science website, http://ecology.cwru.edu/ohio-science/state-belletini.asp (accessed July 6, 2002).

302. This comment from June 14, 2005 can be found at http://www.pandasthumb.org/archives/2005/06/a_new_recruit.html - c35130 (accessed September 9, 2006).

303. This comment from August 4, 2005 can be found at http://pharyngula.org/index/weblog/comments/perspective/ (accessed September 9, 2006).

INDEX

A

B

C

D

E

F

G

H

I

J

K

L

M

N

Printed in the United States
67365LVS00004B/73-75

9 780979 014109